SUSTAINABILITY OF AGROECOSYSTEMS

Edited by **Alexandre Bosco de Oliveira**

Sustainability of Agroecosystems

http://dx.doi.org/10.5772/intechopen.70964
Edited by Alexandre Bosco de Oliveira

Contributors

Hortensia Brito-Vega, Juan-Pablo Martinez-Davila, Lorena Casanova-Perez, Ram Prakash Yadav, Anna Gajda, Karolina Furtak, Matthew Mariola

First published in London, United Kingdom, 2018 by IntechOpen
IntechOpen is the global imprint of INTECHOPEN LIMITED, registered in England and Wales, registration number: 11086078, The Shard, 25th floor, 32 London Bridge Street
London, SE19SG – United Kingdom
Printed in Croatia

British Library Cataloguing-in-Publication Data
A catalogue record for this book is available from the British Library

Additional hard copies can be obtained from orders@intechopen.com

Sustainability of Agroecosystems, Edited by Alexandre Bosco de Oliveira
p. cm.
Print ISBN 978-1-78923-610-1
Online ISBN 978-1-78923-611-8

Meet the editor

Dr. Bosco de Oliveira is an agronomist with a keen interest in crop environment interactions and propagation, germination ecophysiology, and tolerance to abiotic stress in plants, especially drought and salinity. He received his BSc degree in Agronomy (2006) and is a licentiate in Plant Physiology and Horticulture for the middle and high school (2009). Additionally, he received his MSc degree (2008) and PhD degree (2010) in Agronomy/Crop Science from the Federal University of Ceará (UFC, Brazil) and worked as a visiting researcher at the University of Florida (2016–2017). In 2012, he started his current position as an assistant professor at UFC, Brazil, where he has taught courses related to growth and physiology of tropical crops. Since then, he is the head of the Crop Physiology Laboratory and the leader of the Ecophysiology of Semiarid Agroecosystems Research Group, advising undergraduate and graduate students. So far, he published 5 books, 21 chapters in books, and 53 articles in scientific specialized journals, as well as contributed as a reviewer and a member of publisher and various high-impact journals' editorial boards. Due to his relevant achievements in his field, he has been awarded a Postdoctoral Fellowship by the Coordination for the Improvement of Higher Education Personnel (CAPES, Brazil, from July 2016 to June 2017) and a Research Productivity Fellowship by the National Council for Scientific and Technological Development (CNPq, Brazil, from March 2018 until the current date).

Contents

Preface

Sustainable agriculture is a principle encompassing a wide range of environmental, sociopolitical, economic, and cultural issues in rural communities. As a consequence, the sustainability of agroecosystems should not be highlighted as an additional requirement but an overarching concept, which governs the developmental processes on farms and rural communities worldwide. Thereupon, in the pursuit of more sustainable agricultural activities, we all need a clear understanding of the ways—environmental, social, and economic—on how humankind interacts with the earth.

The principles of sustainable agricultural development consist of many components and make up a totality of requirements that are based on the premise of communities having access to and control of local resources, such as seeds, land, and water, as well as working toward local food sovereignty. Thus, to achieve such ideal development in agriculture-related activities, the farmers should be aware of the relevance of the levels of biodiversity and organic matter, the soil health management, and the ability to locally close the biogeochemical cycles involved in agroecosystems. It is a crucial long-term process because the consequences of unsustainable human activities may significantly affect the harmony of social, economic, and ecological equilibrium, threatening the survival of future generations. Hence, both developed and developing countries face gigantic challenges to drive toward sustainability, but as many of these challenges are common to several agroecosystems worldwide, we can share our best practices, principles, concepts, and case studies and, therefore, learn from each other through books such as this one.

In general, this extension and research book explores how production systems can remain profitable while conserving the natural resources and protecting the environment, thus focusing on the man-environment interaction and its role in sustainable agroecosystems. In other words, it highlights the impacts of management practices on crop production, soil health, and environmental conservation. Furthermore, regarding this project's socioeconomic, historical, and political context, it also provides valuable information about managing income-generating rural activities and its effects on humankind's development and cultivation of processes for producing food, feed, fiber, fuel, and other goods.

A successful management of agroecosystems depends on a clear understanding of the "sustainability" term and its related concepts, which will allow the agricultural development to meet the present needs without compromising the ability of future generations to meet their own needs. In order to clarify this point, the present book is composed of modern theoretical and applied studies that highlight the core principles and evidence of a sustainable agriculture. This work is systematically divided into two sections, which summarize crucial insights into this theme, such as agroecological concepts, case studies, soil health, and agroforestry systems. Hence, the literature presented here is a comprehensive overview of current concepts related to the sustainability of agroecosystems, with the main focus on agroecological and integrated farming system models. Such approaches are not only reliable ways of obtaining fairly high-crop yield with a significant scope for resource recycling but also the concept of ecological soundness leading to sustainable development in agricultural lands.

This is definitely a valuable reference book on the sustainability of agroecosystems, particularly for those who work in research organizations and higher academic institutions. It will be of great interest, especially, to graduates, postgraduates, and researchers who work with agriculture, forestry, rural extension, farm development, economy, social sciences, and related subjects. Moreover, it provides a reasonable resource for readers who are interested in a quick review of trending topics about sustainability.

This project includes scholarly contributions by various authors pertinent to agricultural and biological sciences. Each contribution comes as a separate chapter complete in itself but directly related to the book's topics and objectives. Moreover, the chapters included in this book have been written by researchers whose expertise allows the relatively complex sustainable agroecosystem-related topics to be easily understood by any reader. Therefore, the target audience comprises not only scholars and specialists in the field but also common people and enthusiasts about this theme. Such chapter's collection is certainly a valuable resource about agricultural sustainable principles and a pleasure reading for those who are willing to dive more deeply into the study of "sustainability of agroecosystems."

As an editor of this book, I am grateful to all the authors who have written their chapters meticulously and contributed their valuable work in this book. I would also like to thank the editorial staff of InTechOpen Publishing and its team, for all the kind support they have provided me throughout the whole editorship process, enabling the production of this book on time and in a great manner. I express my special thanks to my mother Francisca, my wife Maria, and my kids Matheus and Giovana, for inspiring me and being my pillars of strength. Last but not least, my deepest gratitude is for my Lord and Savior, Jesus Christ, who gives me good health to conduct scientific research and agricultural extension projects, as well as teach and supervise my students. As a conclusion, I have to say that I could not be happier with the development of this book, especially because this idea came up from such a noble activity, which is my passion and has everything to do with sustainability, that is, teaching. As Confucius once said, "If you think in terms of a year, plant a seed; if in terms of ten years, plant trees; if in terms of 100 years, teach the people."

Alexandre Bosco de Oliveira
Professor of Agriculture
Plant Science Department
Federal University of Ceará
Fortaleza, Ceará, Brazil

Section 1

Agroecological Concepts and Studies

Chapter 1

Epistemic and Conceptual Orphanhood in the Sustainability of Agroecosystems

Juan Pablo Martínez-Dávila and
Lorena Casanova-Pérez

Additional information is available at the end of the chapter

http://dx.doi.org/10.5772/intechopen.74110

Abstract

The main problem in the analysis of sustainability of agroecosystems is precisely the lack of definition of both concepts. Above all agroecology has used the concept of agroecosystem in a nominal, ahistorical and atheoretical way; when it is possible to observe it as a system and at the frontier of complex systems. Thus, agroecology does not use, for example, the Thermodynamics of non-equilibrium, Theory of Catastrophes, Complex Networks, Theory of Chaos, among others. It is time to prompt agroecology agroecosystems out of the darkness of closed systems and bring them to the frontier of science systems. The objective of this document is precisely to make the field of agroecological studies expand towards the use of the theory of autopoietic social systems.

Keywords: sustainability, agroecosystem, complex systems, Episteme and doxa

1. Introduction

The concepts of Agroecosystem (AES) and Sustainability were born orphans of a science, because even though Agroecology proposed the agroecosystem as a unit of study and built it considering that the concept Agro contains agriculture (it does not generally study agriculture in its entirety, it does not consider farmers or their management, without involve market, or social consumption), that Eco constitutes an ecological perspective of agriculture in terms of its structure and the relationships between its parts and the system, even though this idea does not contain in its theoretical tools- methodological those offered by systems theory and in none of its four generations of change, was never located epistemologically, or theoretically founded and therefore conceptualized without extension and intension.

The concept of AES, still diffuse, led from the seventies to the eighties the study and transformation of agriculture with good prospects; unfortunately international and transnational organizations with support from the government of economically and politically dominant countries -among many other institutions- introduced the concept of sustainability, it is intuited that this was developed, more in order to divert attention over the catastrophe provoqued of exploitation of the resources around the world. This showed the tacit acceptance that something failed in the design and operation of the neoliberal development model. Related to the above, we can see that even though the objective of conceptualizing the AES required to clarify its structure and functioning, but above all its objectives. This did not happen and today, despite the prostitution of its use and the epistemological, theoretical-conceptual and methodological lack of definition, the term sustainability represents what was wanted: nothing.

In reference to the above, Luhmann [1], at the end of his life, took a pessimistic position with respect to the manipulation that the neoliberal model has promoted in a marketing way towards the individuation of consciousness and functional systems (human beings, institutions, companies, organizations and countries) in such a way that group spirit, solidarity and social compassion have practically disappeared. In perhaps his last writing, he hinted at his position regarding the impossibility of achieving something that could be called sustainability. Luhmann [1] wrote the following: "… Today the problem is much worse than before. We can continue with our habits and return to moral demands, which will be as justified as ever, but who will listen to these complaints and who will react to them if society cannot control itself? And what can we expect if we know that the same success (in economic and financial efficiency) of the functional systems (Institutions, companies, sectors and governments) depends on their indifference towards the social and environmental problems of the world? How could we expect the inclusion of all kinds of concerns within the great social system?"

Therefore, the purpose of this document is to help agroecology, as a science, to expand its epistemological, theoretical and conceptual perspectives, towards the approach of the theory of social systems of Luhmann [2] and the second-order, cybernetics of Heinz Von Foerster, backed by the quantum perspective of Ervin Laszlo.

2. The problematic

From an agroecological point of view and even when the knowledge it generates must be valued positively, it is operated with the type of closed systems, where the influence of what happens in the higher systems is not observed. A simple example can be considered, if in the coffee growing areas of Brazil there was an erratic frost, the grain volume would be lower in the world and its price would grow; all this could generate more income for intermediaries and coffee growers in the regions dedicated to this activity in the state of Veracruz (Mexico) without having done anything to achieve it.

Thus, in general, agroecology does not observe the farmer or peasant, as the cybernetic controller of the properties it manages and this prevents considering that the size of the plot (represented by an agroecosystem) is the size of its decision making capacity (per share). Or omission). This, if you want to investigate the management of the plot and the controller's reactions to the economic, political, cultural and social interferences of their superior systems. In the 'systemic thinking' approach, we are trying to think and do science beyond reductionism (closed systems) according to Casanova et al. [3] allowing us to evolve towards other more encompassing forms that account for social complexity.

Even from the point of view of systems theory (remember that the AES is a system before anything else) there are aporias of the systems approach that it is necessary to recognize. Gharajedaghi in Systemic Thinking: Walking the change or change the way of Herscher [4] states that "*… a synopsis of the main theoretical traditions reveals that, while the analytical approach remained essentially intact for almost four hundred years, systemic thinking has already passed by three different generations of change (today the fourth already dominates).*

"*The first generation of systemic thinking (that of operational research) dealt with interdependence, in the context of mechanical (deterministic) systems. The second generation of systemic thinking (that of cybernetics and open systems) dealt with the double challenge of interdependence and self-organization, in the context of living systems. The third generation of systems thinking responds to the triple challenge of interdependence, self-organization and freedom of choice, in the context of socio-cultural systems.*" The fourth generation still did not appear clearly, but this refers to the study of complexity in systems, which is where systems theory transcends a social vision, with other languages and methods (*Luhmann, Spencer Brown's* Theory of Form, *Heinz Foerster, Aldo Mascareño, and Carlos Maldonado*, among others). This makes it one of the main contributors to the frontier in science, along with neurosciences and quantum mechanics.

To begin, we must consider the principles of Bertalanffy's Systems Theory [5] (although is recognized the Aristotelian fundamental contribution that "*The whole is more than the sum of the parts*"). According to Arnold and Osorio [6], the perspective of the TGS It arises in response to the exhaustion and inapplicability of the analytic-reductionist approaches and their mechanistic-causal principles. It follows that the key principle on which the TGS is based is the notion of the organic totality, while the previous paradigm was based on an inorganic image of the world.

As a scientific paradigm, the *General Systems Theory* (TGS) is characterized by its holistic and integrating perspective, where what is important are the relational functions and the superior systems that emerge from them. As a practice, the TGS offers a technically adequate environment for the interrelation and optimal communication between specialists and specialties, thanks to the construction of a common language. Under the criteria of Arnold and Osorio [6], in a broad sense: "*… The General Systems Theory is presented as a systematic and scientific way of approaching and representing reality and, at the same time, as an orientation towards a stimulating practice for transdisciplinary work forms.*"

Arnold and Osorio [6] define, nevertheless, that: "*… it is convenient to notice that despite its renovating role for classical science, the TGS has not yet managed to influence -in the fundamental- the Cartesian mode (subject / object separation).*" Thus, part of their problems is both the definition of the status of reality of their objects and their concepts, which is now debatable according to

Arnold and Osorio, as the development of an analytical tool suitable for the linear treatment of systemic behavior (causality scheme).

One aspect of high relevance is the location of systems theory in general epistemology. In this respect, Arnold and Osorio [6] consider that systems epistemology refers to the distance of the TGS with respect to positivism or logical empiricism. On the other hand, Bertalanffy, referring to himself, says: "In philosophy, the author's formation followed the tradition of the neopositivism of the Moritz Schlick group" later denominated The Vienna Circle, but his interest in German mysticism, historical relativism of Spengler and the history of art, together with other unorthodox attitudes, prevented him from becoming a good positivist [6]. This is a good advantage for agroecology, considering that the Theory of Social Systems Autopoietic (Border in the study of the Systems) has not yet managed to integrate the materialistic conflict and contradiction into its ideology.

Finally, agroecology has not been able or has not wanted to reinforce itself epistemologically and this is counterbalanced from the Hermeneutics, with adequate conceptual interpretations. In this regard Goode and Hatt [7] pose. "Science abstracts from reality and examines certain aspects of phenomena and not the totality of the phenomena themselves. In truth, separating any phenomenon from that with which it is related constitutes a fact of abstraction. Since science attempts to investigate certain sectors or aspects of reality, using them to interpret them as an abstract system of thought, it should not be surprising that, in order to communicate their findings, each of the sciences uses terms or concepts that are their own. Now, we use these concepts to represent the phenomena or aspects of them, which we are investigating. Therefore, when a proposition is formulated, the concepts are used as symbols of the phenomena being studied. However, because we are dealing directly with only the concepts, it is clear that the concept can be confused with the phenomenon of what is supposed to be a symbol."

This is a common mistake named objectification. It is often forgotten that concepts are logical constructions created from impressions of the senses, perceptions, and even complex experiences. The tendency to assume that concepts actually exist has led to many failures. The concept is not the phenomenon itself; that is, these logical constructions do not exist outside the established frame of reference. The incapacity to recognize this difference is what has been called the fallacy of objectification, that is, abstractions are treated as if they were phenomena. Finally, we will say that facts as concepts are abstractions then they have only meaning in a frame of reference within some theoretical system, Goode and Hatt [7].

We can now understand that the agroecosystem is a concept and is not a plot under cultivation. We must understand that the agroecosystem is a representation, of a cut of the agricultural reality, it is a model, an abstraction that the scientists create to understand the agricultural reality (See **Figure 1**).

The notion of doxa is as old as Greek culture, but it was Plato who gave structure to the epistemological comparison between science and opinion; which is inspired by the myth of the cave. Plato accepts that the human being is intelligible and sensitive, but that if he cannot leave the cave, his reality is proportional to what he sees (body, desires, imagination in the shadows he observes on the wall). If he does not have the capacity to think, he can only have will. But if you leave the cave, learn to think and probably come to have beliefs through the

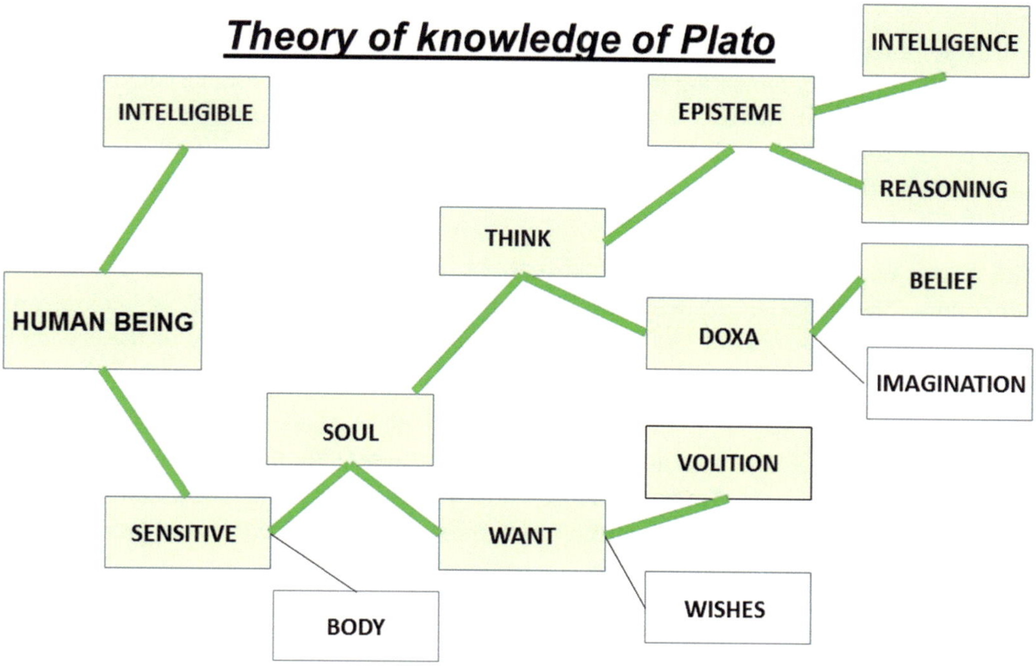

Figure 1. Theory of knowledge of Plato. http://historiadelafilosofiaparacavernicolas.blogspot.mx/2015/ [8].

DOXA (*The Reality, The everyday*)	EPISTEME (*Science*)
• Nature;	• Ecosystem;
• Happiness;	• Sustainability;
• Electric System;	• Plausible System;
• Mechanic System;	• Plausible System;
• The Agricultural;	• Agroecosystem;
• Installation;	• System;

Table 1. Distinction between philosophy of science and orthodoxy.

doxa. When the human being learns to think and creates epistemology, based on it, he can do philosophy and science.

The agroecosystem, as can be seen in **Table 1**, is a representation, a scientific model of a doxic reality. Inclusive, the words or the doxical terms, when using them in the dialog, in the writing, in the poetry are only symbols of real objects. When a message is written that says: "*Please stow the chair that is in the yard*" the word chair is not a real object, it is a symbol of the real chair, if it should not have been sent wrapped in the message the chair in discussion. If I say that I bought four melons: Would I have to bring them in my hand when I referred to them? Or I tell my brother who lives in Canada that I have a new girlfriend: Should I send her to Canada before talking on the phone with him?

A clear example of the above is the orthodox word, is an adjective that means that it complies with traditional and generalized norms in an institutional or regional (cultural) scope is something legitimate, something right or true, which is followed by most of a community, but it is not scientific. Heterodox is something that is not orthodox, therefore, it is something little used by the community. A heterodox is someone who is not satisfied with the dogma and beliefs of a particular religion, or with the ideas or practices of a cultural doctrine and generally accepted, but that is not science either.

This is one of the main problems in the training of researchers in the sciences related to agriculture (Natural Sciences) that affects their relationship with General Epistemology and impedes them, due to their super specialization from being located in any of the epistemic traditions, in some of the currents, either Galilean or Aristotelian.

In the first instance, integrating this knowledge allows the researcher to create mental structures, to locate himself in a precise epistemological line, to identify the theory or theories that best contribute to understand, interpret or explain the phenomenon that will be studied. In this line of thought we can build really plausible hypotheses, design methodological processes that will be practically "theory in action."

3. Epistemical and conceptual analysis

From the epistemological perspective we could rely on Mardones and Ursua [9], who begin their reflection with the following: "*... If we look at the panorama of the philosophy of science, or reflection on science and what it has that to be considered as such, from the height of its history, two important traditions are distinguished; the Aristotelian call and the so-called galileana.*" They are two perspectives of science or two different approaches about the conditions that an explanation that wants to be called scientific has to satisfy. Both traditions have their roots and representatives in the Greek world. In order to locate the agroecosystems in an epistemological position, we must base on which of the two has its origin.

Mardones and Ursua [9] state that: "*Nothing happens in the cultural and human world overnight. The ideas are incubating slowly or more rapidly, under the influence of social, political, economic or religious events.*" And they add that the conception of the world, fruit of the new way of looking at it, which is already visible in men like Galileo or Bacon, is not so metaphysical and finalist, as *functional* and *mechanistic*. The new eyes of modern science are eager for power and control over nature, society and the economy.

Thereby, the center is no longer the world, but the individuation: the human being. For this reason his empiricist look reduces in object to the human being and nature for his needs and utilities. Thus, in argues of Mardones [9]: "*This pragmatic interest, mechanic-causalist, that is not going to ask anymore already by the why (Aristotelian)" and "for what" last, but by the "how" more immediate and practical of the phenomena and their consequences, interest that emerge with force in the century that goes from 1543, the year of the appearance of the work of Copernicus "De revolutionibus orbium coelestium," until 1638, date in which see the light the "Discorsi" of Galileo.*" According to Mardones: It will be Galileo then, a typical representative of the new mentality that changes the qualitative physical explanations of Aristotle, by the mathematical formulations of Archimedes.

In the arguments of Mardones and Ursua [9], *the beginning of mass production, according to the scheme of supply and demand favored the accumulation of capital and the strengthening of a new urban socio-political class: the bourgeoisie and the Bourgeois State. "Characteristic of this social class will be the taste for a more secular culture, a propensity for concrete actions and their sense of order and the positive." According to Comte, the positive, which is what our popular language has collected in expressions such as "**go to the positive**," that is, **to the useful and pragmatic**. The new science gathers this objectivist interest, in accordance with the attempt to dominate the human being and nature, pointing out a technological attitude of knowledge* and its applications.

The "new science," according to Mardones [9], which replaces the science Aristotelian, *will consider as scientific explanation of a fact that which comes formulated in terms of laws that relate phenomena determined numerically, that is, mathematically.* These explanations will take the form of causal hypotheses, but causally it will have here a functional connotation in a mechanistic perspective. The touchstone of the value of causal hypotheses will be determined by experimental analysis. It will be the comparison of the hypothesis with the consequences deduced through the observation of reality or experimentation which will tell us an explanatory value, the human, social and cultural, the historical-critical, cannot be scientific, it is an art, like physicists will say.

In an acute synthesis, Mardones and Ursua [9] define: "*… We will now understand why, speaking in a very broad sense, the confrontation can be expressed in general terms and this can be expressed in terms of causal explanation versus teleological understanding or, as we will say more ahead (Erklären: Explanation) against understanding (Verstehen).*"

3.1. Concepts of sustainability and agroecosystem

The concepts have been—almost always—poorly constructed (especially in agricultural research), there is no intentionality or clarity in its content. As an example, the concept of Agricultural Development (linked to the Agroecosystem concept and contributes since economically perspective to define Rural Development) fails to define what "*development*" is, even though since the [1973] Weitz [10] in Rehovot, Israel, politicians and scientists from all over the world, agreed that any development is "*a process of change.*"

Hence, in the Agricultural Development who changes? O Who should change mainly? Should it be the farmer? Who is the protagonist of that level of development and who must change based on his historical-social determination and the identity of his territorial appropriation. It is now necessary to ask: what should change? Well, mainly in his attitude and behavior; towards what? Logically, towards the management of your plot (Agroecosystem in the design and monitoring evaluation phase) how will it do it? Maybe you could do it in a planned way (Diagnosis, Strategy and Evaluation) and why do all this? In order to fulfill its social responsibility in an optimal way, the land was given, in some cases also the water, the bank financing and perhaps an agricultural insurance, to produce the food, raw materials and environmental benefits that the society and its family demand, but also to improve their quality of life. And who would lead this process? We could say that the Law for Rural Development which defines the municipal authority to coordinate it, although nothing prevents this from being done by an organized group of farmers.

We could now build the concept based on the previous ideas. *"The Territorial Agricultural Development (DAT) is a socially and historically determined process and main contributor in the economy of Rural Development (Major System). The DAT is a process in which through changes in attitude and planned behavior of farmers in the management of AES produce optimally, ecological, economic and social, food, raw materials, quality of life and benefits environmental factors that the market and the population demand. Conducted by Organized Farmer Groups."*

Certainly, the concepts are dynamic and flexible. Other elements of a higher order could be inserted in this conceptual construction, which would give greater strength to the concept and Territorial Agricultural Development; as it can be the introduction of empowerment in the organized group. A brief definition of this empowerment indicates that what is sought first is Social Empowerment (Large and strong organized group); then build an Economic Empowerment (Installation of efficient and capitalized companies) Finally, seek political empowerment, first at the municipal level; later in the state and federal deputation. The latter to keep your project in time without external interference [11–13].

3.1.1. Sustainability

To date, there is a large number of definitions of sustainability, developed without addressing a minimum methodological structure. Generally, there is no major system that contains the concept under construction. Its structure is not described (parts that compose it) nor its function (relations between the parties) except its protagonist and its objectives. The authors in this sense could make a simple definition could be built but with the necessary quality to walk towards a plausible concept.

Should Sustainability be local or global? Nowadays it would seem that the answer is that it is local: My sustainable experiment, the sustainable radishes, and sustainable shoes; in short, today everything is sustainable. Imagine that Australia is totally sustainable (whatever that means), but the rest of the world is not and suddenly (exaggerating) the sea level rises 40 m, flooding its main coastal cities. Australians would say: but why, if we are sustainable. Indeed, they are, but not others. What does this tell us? Well, sustainability must be considered on a global level, if it does not work. There must be dynamic balances between the economic, political, cultural, social and environmental dimensions; depending on the quality of the resources that are managed.

Thus, we can say that: *"Sustainability is a political and global process (upper system) where what must be sustained and support itself is the life of the Human Being and all the living on the planet, through a rational management of social and economic processes and that along with a management, also rational, of nature, the human being ever scope the happiness, whatever that it may."*

3.1.2. Agroecosystem

As it indicated earlier, the AES is first of all an epistemic tool (theoretical-methodological) that we call System, as a model of a part of the empirical reality. Then *"the Agroecosystem is a representation or model of a cut of the agricultural reality, managed by a cybernetic controller (farmer*

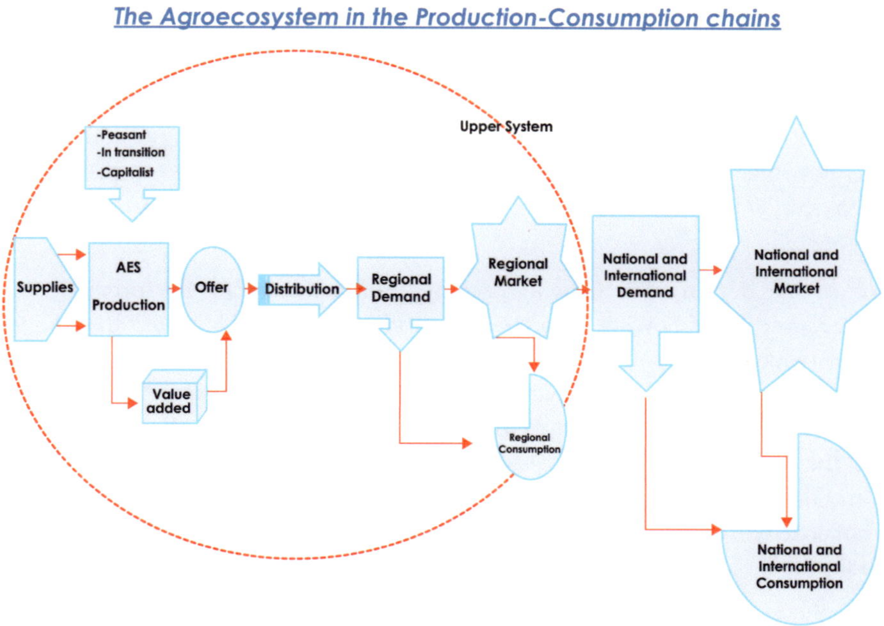

Figure 2. The agroecosystem in the production-consumption chains.

or peasant) to produce and satisfy -in an optimal way- the social needs and demands of the market. Conceptualized as an organized totality (that is why it is a system) in which the elements are not separable and, therefore, should not be studied separable." In addition to the linguistic and epistemological clarity that this approach provides, is that being the AES a model, multiple simulations can be designed in much less time and with lower costs. Unlike the development of controlled experiments in the field that will surely take much more time and financing.

In **Figure 2**. The Agroecosystem (AES) is defined as the initial and main link of the Production-Consumption Chains. In the AES the first level of supply of agricultural production is activated (fresh or with added value) and the distribution processes begin to satisfy the estimated demand, the regional, national or international market and therefore consumption. In this apparent and simple process occurs the most important phenomenon of social reproduction: Consumption - to live - inexorably depend on the AES and in that same process the AES also depends on consumption (family and social) to survive. This constitutes what the higher system of the AES may be committed to by its social responsibility: Feeding society.

4. A new agroecology

With everything and how valuable Agroecology is, it certainly should not remain static and enclosed in its box forever. The agroecological leaders in the world could perhaps be integrated to new approaches to study the Agroecosystem. One of them is that of complex systems with cybernetics of second order and therefore with researchers at the observer level of those who observe [Researchers who study farmers observing their agricultural processes (Agroecosystem) has the great advantage that the observer of second order (the researcher) can observe himself in the same observational process].

Criticism about the biological-ecological approach is by no means new, since 1980 Dr. Efraím Hernández Xolocotzi [14] stated that the principles derived from ecology should be checked, as soon as possible, without forgetting: *That the essence of agroecosystems, is the management of natural resources by man and that studies with an ecological focus, have disparaged, precisely, the presence of man.*

Therefore in the design of a new agroecology should consider the previous advice of Hernández X. and the following theoretical criticisms. Muench [15] stated that the tendency that has predominated over the interpretation of the Agroecosystem is that which considers the technique abstracted from the whole system of social relations; performing an isolated examination of it and establishing a rigid relationship of cause and effect between the technique and the socio-economic instances of society; defining the technique as an independent variable and the economy as a dependent variable. Thus they come to the conclusion (although they do not specify it) that social systems are determined entirely by changes in technique and need to adapt to it.

Faced with the avalanche of criticism that progressive scholars make to the aforementioned tendency, Muench [15] himself expresses his position about it: "*... Before the development of irrational forms in the production of satisfactions carried out by the tendency described above, arises, as a weak humanist defense of resources, another tendency of the conceptions of the scientific research of agriculture. This tendency is born from the field of natural sciences, specifically from ecology, from where the concepts and basic categories for the method of study are taken.*"

The most important contribution of this conceptual tendency - to say of Muench - lies in the fact of demanding an integral analysis of the agricultural phenomenon; however, an important basic error is made: a social phenomenon - agriculture - is analyzed through a method proposed by biology and ecology. Taking as a fundamental concept the ecosystem and applying the method of study designed for the study of natural ecosystems, for which the concept of agroecosystem is elaborated, which is nothing else than an ecosystem where man intervenes as one more organism in the trophic chain, Muench [15].

Pablo Muench [15] concluded: "*Obviously this conception leads to an analysis of agriculture outside the historical development of society, without making a clear distinction between human work and the other elements of the production process; it does not assume the importance of the unequal development of agricultural processes with different social production objectives; they interpret the efficiency of agricultural systems apart from concrete social conditions and deny that production systems have greater material conditions, developed by society in their historical process, than the conditions imposed by nature. To consider agriculture in that way is indisputably wrong.*

In addition to the above, the concept of Agroecosystem was managed without constructed it, even methodologically, and logically this has led to diagnose it without knowing it, to design it and operate it without theoretically knowing its structure and operation, and finally to evaluate it without applying the minimum elements of the theory of systems.

Regarding the reality status of systems and therefore of agroecosystems, the fourth generation of systems, that of complex systems, introduces a highly clarifying idea for a potential new agroecology: Systems are not things, but those things we see them or rather we want to see them as systems. That is, we use the filter of systems theory to see reality, not because systems exist in reality, but because we want to see it through that filter, which is epistemic and therefore theoretical; as we could see reality with any other filter, with a positivist or dialectical approach, perhaps religious -if we would like to do so- in recent times it is said that reality is not objective, it is undoubtedly an interpretation of the human brain, there in his dark room, guided only by the deceptive senses.

With regard to the above, Herrscher [4] President of the International Society for the Systems Sciences, the leading entity of systemic thought in the world, receives a basic question: What is a system? And he replies: *"Almost everyone will tell you that it is a set of elements interrelated with a common goal … But in reality we are the ones who do it system: when looking at it, seeing it as a system, when deciding to consider it as a member of a category that has certain properties marked in his theory." His interlocutor answers him with another question - Is it to say that all things in this complex world are systems? - Properly Herrscher points out that: "… 1) systems are not 'things', but there are things that we decided to treat as systems, and 2) that not even all of them are 'things'. There are artifacts like a car or a plate that we call mechanical systems. There are biological systems: living organisms like the dog and the cat that we have at home, or like each one of us or, specifically, our bodies. There are organizations like your factory, or like our families, or like our government or our country (which is not the same) that we call social systems. And there are systems of ideas, beliefs or behaviors, such as ideologies, religions, and cultures."* Science is correcting itself, because we see here that Herrscher is still considering doxical elements as epistemic (mechanical systems, cats, dogs, etc.).

Herrscher [4] concludes by saying that *"The condition of the system is not an intrinsic quality of the thing - but an attitude or appreciation of each one. For those who consider, based on a theoretical approach, that in reality there are things that are related to one another that tend to the formation of a human being in an articulated and meaningful way it is a system."*

Garcia [16] in his classic book "Sistemas Complejos" confirms what Herrscher said about the reality status of systems as an object of study, he says that: *"A complex system (AES) is a representation of a cut of that reality, conceptualized as an organized totality (hence the denomination of system), in which the elements are not 'separable' and, therefore, should not be studied in insolated."*

In other words, a new agroecology, before studying natural resources and food production must understand and interpret the culture of the social group that wishes to study. To understand their relationship rules, not break them untimely. If the producers did not ask for technical help, it is not that they do not need it, they probably do not want strangers in their locality and they consider themselves assaulted by the presence of individuals, who even arrive with a haughty and arrogant tone.

Do they really want advice of a productive and ecological nature? Probably what they want, although they do not manifest it, is the exploitation of non-agricultural materials. Like lime, marble, stone or sand, circumstances that many times the agroecologist does not see or does not care about. Better still, is to listen unattended demands, such as small dams, roads, electricity or drinking water; demands that could seek the solution to technical or financial problems.

Remind that the problems of farmers, in reality are elements that prevent achieving their objectives. And that those problems and objectives have an owner and these are the farmers. Our problems and objectives should not overcome theirs.

5. Conclusions

In spite of how hard epistemological criticism can be, it must be recognized that the scientific activities developed by Agroecology are of high biological and ecological value. However, if this science does not come out of its closed box and does not attend higher level phenomena, which undoubtedly affect the behavior of Agroecosystems and its Cybernetic Controller, in order to fulfill its social responsibility, it is most likely that AES be considered by other professionals in the dynamics of complex autopoietic systems and be mounted on the frontier of complexity sciences.

When agroecologists or scholars of agroecosystems begin to understand and clarify previously diffuse meanings, agricultural research will advance, not only in knowledge, but also in the dynamic equilibrium of rural activities. But above all in the philosophy necessary to generate new and renewed scientific knowledge that contributes to achieving economic, social (with its political and cultural manifestations) and environmental well-being.

Just to introduce us to the topic, the semantics of the Theory of Social Autopoietic Systems of Niklas Luhmann considers that we have been wrong to consider the Society as if it were the Population. It should be noted that at the beginning of each social group, complexes of relationships were created in the processes of appropriation and adaptation of the groups to their environment, which also built identity. This complex of relationships, as the generations pass, becomes a collective memory and then a social memory. That social memory in this historical moment is no longer concrete but abstract, imbued in the processes of identity-and although it is now abstract-it determines the behavior of social groups. The social and collective memory is reproduced in time (it is autopoietic) that generates a culture that translates into specific traditions and behaviors. When it matures and is regulated, this social memory (Culture) generates Civilization and Society (orderly social relations as in the social division of labor), which is reproduced in time and based on it individuals, institutions, companies, governments and countries build its autopoiesis, and although it is maintained in abstract conditions, the traditions and concrete behaviors, also remain as collective memory.

Author details

Juan Pablo Martínez-Dávila[1]* and Lorena Casanova-Pérez[2]

*Address all correspondence to: jpabast@hotmail.com

1 Colegio de Postgraduados, Campus Veracruz, Veracruz, México

2 Evaluation of Sustainable Programs for Certification, Veracruz, México

References

[1] Luhmann. Globalization or World society: How to conceive of modern society? International Review of Sociology. 1997;**7**(1):67-79

[2] Luhmann. Introduction to the Theory of Social Systems. Book, ISBN: 968-859-318-4

[3] Casanova PL, Martínez DJP, Ortiz SL, Sánchez CL, Romero GL, Olvera BP. El agroecosistema comprendido desde la teoría de sistemas sociales autopoiéticos. Revista mexicana de ciencias agrícolas. 2015;**6**(4):855-865

[4] Herscher E. Pensamiento sistémico. Buenos Aires, Argentina: Ediciones Granica S.A.; 2003

[5] Von Bertalanffy L. Teoría General de los Sistemas. México: Editorial Fondo de Cultura Económica; 1976. p. 336. ISBN: 968 16 0627 2

[6] Arnold M, Osorio F. Introducción a los conceptos básicos de la teoría general de sistemas. Cinta de Moebio. 1998;**3**:40-49

[7] Goode W, Hatt P. Métodos de Investigación Social. México, D.F.: Editorial Trillas S.A; 2002

[8] http://historiadelafilosofiaparacavernicolas.blogspot.mx/2015/

[9] Mardones JM, Ursua N. Filosofía de las Ciencias Sociales y Humanas. México, D.F.: Ediciones Coyoacán S.A. de C.V; 2003

[10] Weitz R. De Campesino a Agricultor. México, D.F.: Fondo de Cultura Económica; 1973

[11] Rappo S. Agricultura orgánica y comercio justo: estrategia de sobrevivencia para los pequeños productores rurales. Aportes, Revista de la Facultad de Economía, BUAP. 2009;**XIV**(40):59-81

[12] Gutiérrez H. Ficha final de la experiencia: Sociedad Cooperativa Agropecuaria Regional *Tosepan Titataniske (Unidos Venceremos)*. Cuetzalan, Puebla, México: Observatorio de Soberanía alimentaria y Agroecología; 2011

[13] Diego N. Informe de la Sociedad Cooperativa Agropecuaria Regional (Tosepan Titataniske). Cuetzalan, Puebla; 2006

[14] Hernández E. La producción de alimentos básicos en Yucatán. en Seminario sobre producción agrícola en Yucatán. Gobierno del Estado de Yucatán. Secretaría de Programación y Presupuesto, Secretaría de Agricultura y Recursos Hidráulicos, Colegio de Postgraduados en Chapingo. Mérida, Yucatán; 1980

[15] Muench P. El concepto: Proceso de Producción Agrícola. Gobierno del Estado de Yucatán. Secretaría de Programación y Presupuesto, Secretaría de Agricultura y Recursos Hidráulicos, Colegio de Postgraduados en Chapingo. Mérida, Yucatán; 1980

[16] García R. Sistemas complejos. Conceptos, método y Fundamentación epistemológica de la investigación interdisciplinaria. Barcelona: Gedisa; 2006

Chapter 2

Limited Fertility, Limited Land: Barriers to Sustainability in a Chilean Agrarian Community

Matt J. Mariola

Additional information is available at the end of the chapter

http://dx.doi.org/10.5772/intechopen.75210

Abstract

There is an ongoing debate within the field of agroecology about the importance of socio-economic context as a causal factor of agricultural unsustainability. The ultimate thesis of this paper is that the causes of unsustainability cannot be fully grasped without understanding historical trajectories of land use and land tenure. I investigate a community of indigenous smallholders in southern Chile who have expressed a desire to find an ecological alternative to the input-intensive commodity cropping system they currently practice. What factors motivate them, and what are the key barriers holding them back? Interviews with 38 individuals and 3.5 months of participant observation reveal a complex causal chain. A key barrier to an alternative, more sustainable agricultural system is a deficiency of organic animal manure, of the animals which produce it, and of the land base on which to sustain sufficient numbers of livestock. But these factors can only be understood in the context of the historical dispossession of Mapuche territory by the Chilean state over a century ago, which led to socioeconomic marginalization by pushing them onto the poor soils and steep hillsides they currently occupy. I close by emphasizing that the path to agricultural sustainability must combine both sociopolitical and agronomic considerations. Land access and land tenure are as important to achieving sustainability as land management.

Keywords: Chile, land tenure, manure, Mapuche, organic, political agroecology, repeasantization, soil fertility

1. Introduction

Prior to the arrival of Spanish colonists in the sixteenth century, the modern-day country of Chile was inhabited by a variety of indigenous groups stretching from the arid deserts in the north to windswept Patagonia in the south. The Spanish arrival resulted in the rapid deterioration of indigenous populations via genocide, illness, and forced displacement. One group, however, famously resisted the Spanish and retained a large autonomous territory in the south even after Chile established independence from Spain: the Mapuche, called *araucos* by the Chileans, namesake of the present-day region of Araucanía.

This autonomy ended in the mid-1800s, when the Chilean government embarked on a military campaign to "pacify" a people they deemed an impediment to the advance of civilization and modernity. A sequence of military incursions eventually defeated the Mapuche, a population that had numbered around 1 million at the time of first Spanish contact now reduced to some 150,000 [1]. It is what happened after pacification, however, that had the most profound impact on succeeding generations of Mapuche.

The Chilean state wanted to encourage agricultural settlement of the fertile but densely wooded Mapuche homeland, so they encouraged Chilean and immigrant colonists (*colonos*) to push south, clear the forests, and establish homesteads. A Chilean writer in 1848 described the "phalanx of peaceful immigrants, of industrious settlers" establishing farms in the south ([1], p. 94), while a Dutch settler is quoted, "Each family received a pair of oxen and a cart, an ax, and a rifle to defend themselves from the indigenous people" ([2], p. 239). Between the 1850s and the 1870s, an estimated 14,000 *colonos* settled in Mapuche territory ([1], pp. 94–95). By the end of the century a travel writer could note the extensive clearing of land: "I rode through green fields with stumps scattered through them, all the way from Concepción to Temuco. Men were cutting farms out of the woods, and here and there the wheat was growing among the burned timber" ([3], p. 90).

The Mapuche were deeded parcels by the state, but "unscrupulous land-grabbing at the expense of the Mapuche was an altogether predictable feature, and remained so for the next few decades" ([1], p. 96). Meanwhile, the government set up the equivalent of North American reservations (*reducciónes*) for the Mapuche, diminishing their historic territory by 95 percent ([2], p. 46). Communities were forced east to the mountains and west to the marginal lands of the Pacific coast, where, no longer able to migrate across a large territory with their animals, they took up farming on small household plots. When Green Revolution practices (commercial seed varieties; synthetic fertilizers; chemical pesticides) arrived and seemed to promise an immediate increase in yields, they were promoted heavily by the government and adopted enthusiastically by rural communities – particularly the cultivation of commercial potato varieties as a commodity crop for primary income generation.

The site of my fieldwork is one such community, called Llaguepulli. With the exception of two individuals who have married in, the entire population of Llaguepulli is Mapuche. Many of its residents have migrated away in recent decades, seeking employment in the regional or national capital. Of those that remain, nearly everyone lives the life of an agrarian smallholder: cultivating home gardens, tending small flocks of animals, and growing staple crops on small acreage using conventional, agrochemical-dependent methods.

Llaguepulli is rural, poor, and resource-deprived; in that sense it is like every other Mapuche community in the region [4]. But in another sense it is unique: its main leaders as well as a large portion of its members have expressed an explicit interest in finding an alternative agricultural path to the high-input, low-value commodity crop system in which they are currently immersed. With a progressive leadership and the support of a community-based development NGO they have made progress, including the establishment of a culturally-oriented tourism program, an agroforestry project, and a savings and microloan program run entirely by community members [5]. Yet nearly all the farmland in the community is devoted to basic commodity crops (chiefly potatoes and wheat) fed with synthetic fertilizers, even as the farmers themselves lament this fact. Many barriers to sustainability remain, and the central inquiry of this paper is what those barriers are.

Hovering in the background of that question is a broader one of significance to the field of agroecology. How carefully do we need to consider social, political, and historical forces – which have traditionally been considered outside the scope of agroecology – in order to understand the sources of an unsustainable agricultural system? My broader thesis is that the biophysical and technical causes of unsustainability that have traditionally been the focus of agroecologists are *interlinked with sociopolitical and historical causes*. One set of forces cannot be understood without the other, for they form an interwoven causal chain. I begin the literature review by elaborating on this thesis.

2. Literature review

I break the review of the literature into two parts. First I will discuss the scholarly debate over the incorporation of social considerations into the field of agroecology. Then I will move to the field of agrarian studies and describe recent discourse on the idea of "repeasantization". Ultimately my intention is to draw a link between the two and argue that a process of resurrecting sustainable peasant livelihoods can only take place by giving sufficient attention to the politics of land access among smallholder and indigenous populations.

2.1. Social considerations in agroecology

Open the pages of many agroecology texts and you will find research that could belong in a journal dedicated to the larger field of ecology. *Agriculture, Ecosystems and Environment*, the highest ranking agroecology journal [6], prioritizes research "at the field, system or landscape level" and "data-based biophysical modeling" focused on "biological and physical characteristics of agroecosystems" and "relationships between agroecosystems and the natural environment". The journal's Aims & Scope page does mention sustainability, global environmental change, and land use, but only in abstract terms, with no mention of humans, societies, economics, or political systems. Agroecology is framed as simply "the application of ecology in agriculture" ([7], p. 2).

On one hand this is unremarkable. A field dedicated to the quest for a more sustainable agriculture must be rooted in the science behind sustainable production techniques. On the other hand this exclusion of social, political, or economic components from the field's top journal is troubling. As scholars we are fundamentally interested in *causality*, and to think of agro-environmental problems only in terms of technical or ecological parameters is to narrow one's vision, for the causes and solutions to problems of agricultural unsustainability are nearly always *rooted in social conditions*. Agriculture is not merely a biophysical endeavor but "a complex socio-ecological system", conceived of and carried out by humans embedded within an economic-political-cultural matrix ([8], p. 6).

The question of what role "the social" should play within the field of agroecology has been under intense discussion for the better part of a decade, with roots going back even further (e.g., [9]). For the first half of its near century-long development as a field, agroecology tended to "emphasize the crops and not the people who grew them" ([10], p. 22), but the social sciences have become increasingly integrated in order to "better understand the complexity of agriculture that emerges from unique sociocultural contexts" ([8], p. 3), from the "wider social formations in which [farmers] are embedded" ([35] p. 12).

The more recent advance is not merely to study social forces, but to advocate for *social justice* for small farmers. Two prominent agroecologists explicitly frame agroecology as a "revolution" whose purpose is to "ensure food sovereignty and empower peasants" [11]. La Vía Campesina ([12], p. 15), a global network comprising the world's largest peasant advocacy movement, does not mince words: "Agroecology is a struggle to liberate ourselves from the global capitalist system and attain food sovereignty and autonomy of the people".

And how is this struggle to be carried out? Put simply, since "the sustainability of an agroecosystem is not just the result of a series of physical and biological properties, but also the reflection of power relations," the "practical dimension of agroecology requires politics" ([13], pp. 46, 50). Without a political component focused on achieving food sovereignty for peasant households, agroecology is "just a technical fix" [14]. This paper does not pretend to identify a comprehensive solution to agricultural unsustainability, but it proceeds from the assumption that socio-political factors must be considered in order to identify both a problem and its remedies.

2.2. Repeasantization

As the Results section will show, what the community leaders and inhabitants of my field site of Llaguepulli desire is to regain some of the autonomy over their farming systems that they traded in when they began following a conventional, input-reliant form of agriculture decades ago. They might do this by converting from conventional to organic practices, by growing alternative crops, by developing new market channels, or some of the other strategies peasant groups around the world are using to find alternatives to the Green Revolution model. They thus hope to join a process which has been given the term "repeasantization".

The scholarly discourse on repeasantization links to earlier debates about "depeasantization," and to a century-long discussion of "the agrarian question" before that. It is voluminous, and I do not aim for a comprehensive treatment; recent works have accomplished this with skill and lucidity (see [15, 16]). Rather my goal is to sketch out the parameters of the topic as they apply to the situation in Llaguepulli. I will structure my discussion around a set of linked questions: What is driving the process of repeasantization; what forms does the process take; and what are the main impediments it faces?

2.2.1. *What is driving the process of repeasantization, and what forms does it take?*

The elements of a formal social movement advocating for repeasantization are powerful and manifest on a global scale. La Vía Campesina has grown from a loose collective of 47 peasant and rural worker organizations in 19 countries in the early 1990s [17] to a network of 164 member organizations in 73 countries today [12], and is "considered by many to be the most important transnational social movement in the world" ([17], p. 150). An ever-expanding body of case studies spanning every continent indicates that a growing number of farmers, villages, cooperatives, and peasant groups is seeking an alternative way to produce, distribute, and consume food (see [18]), and what makes the case compelling is the reasons that underlie these efforts.

The basic agricultural system that peasants have either willingly or been forced to adopt in the second half of the twentieth century has core features that are nearly ubiquitous, regardless of crop or geography: it is "a high external input, fossil fuel based, export oriented, monoculture cropping system" ([10], p. 24). In other words, small farmers the world over switched from a "peasant mode" of agriculture, based on growing a diversity of crops, exchanging them within local markets, and prioritizing food security, to a conventional mode based on growing a limited set of commodity crops for external markets using agrochemicals. This switch weaves smallholders into the fabric of global capitalism, where they are subject to the whims of market forces (via input costs and market prices) and placed at a distinct disadvantage relative to the scale of the agro-industrial complex (competing against large, often state-subsidized commodity growers). "The global profitability of farming activity has been progressively declining … as a consequence of the unequal relationship of exchange between the agrarian sector and the industrial and service sector" ([13], p. 48).

Hence, there is a set of conditions whose effects on peasant livelihoods have accumulated for decades, leading to a shared sense of aggrievement towards the capital-intensive agricultural model. So, to the degree that we are witnessing a repeasantization movement, it is not that peasant agriculture is naturally superior to large-scale capitalist agriculture, but that the industrial system has failed to meet the needs of smallholder households. "When the political-economic arrangements upon which entrepreneurial farming is grounded begin to erode, … peasant agriculture re-emerge[s] again as a far more resilient and economically effective alternative" ([35] p. 13).

Corrado [36] argues that the "new peasant condition" is formulated around three poles: the centrality of food to the peasant lifeway; an economics of self-sufficiency and pluriactivity; and the importance of "territory," by which she refers to both the set of local natural resources (soils, seeds, etc.) as well as localized social relations. These three poles would seem to be backed up by the public face of the movement – for example, La Vía Campesina's most recent annual report makes an impassioned call for food sovereignty, greater economic and political freedoms, and peasants' struggle for territory.Van der Ploeg's book and article in *The Journal of Peasant Studies* are the most thorough attempts to theorize the repeasantization process [16, 35]. He describes six major "avenues" along which repeasantization takes place:

1. *Land* is perhaps the single, central resource of concern for peasants. It is crucial as the basis for both food sovereignty and ecological capital. This has not changed much from peasant struggles of decades ago, but there are now renewed options for the use of that land.
2. *Self-provisioning*, which combines the idea of control over one's own food supply (food sovereignty) with control over the resources needed for the continued (re)production of the household.
3. *Proactive market integration*, meaning not that peasants always immerse themselves fully in circuits of capital, but that they have greater power to choose their degree of integration into or distantiation from markets.
4. A greater emphasis on *agroecology*, or the effort to improve a household's ecological capital, or what Van der Ploeg calls the "co-production of humans and nature".
5. *Proactive resistance* against the agro-industrial complex, not via political measures but via alternative production practices and market networks.
6. *Alternative markets* for the distribution and sale of food.

Van der Ploeg can be accused of being overly optimistic about the potential for peasants to take control of their fields and their fates in the face of powerful countervailing forces, and indeed there are scholars who remain skeptical of the power of a movement framed around food sovereignty [19]. Though Van der Ploeg's aim is not to advocate but to describe the general contours of a varied movement, the skeptical view hints at the need to consider the more critical question of what barriers stand in the way of repeasantization.

2.2.2. What are the main impediments to the repeasantization process?

If it is true that new possibilities exist for peasant producers to reclaim autonomy and use ecological capital to build new rural livelihoods, it is also true that an array of powerful forces run counter to their efforts. For example, many peasant communities are located far from city centers and linked to them via dirt or gravel roads, increasing transaction costs and opportunity costs. In the absence of direct market access, marginalized smallholders' only option for selling their crops may be the one intermediary servicing the community, who can pay low prices due to a lack of any other buyers [20]. As we will see below, in Llaguepulli the intermediary potato buyer acts as a monopoly agent with control over the price offered farmers.

If farmers cannot provide the nutritional needs of their family, and if they cannot generate income by selling their products for lack of market access, they may have little other recourse but to seek out wage labor, either in the agricultural sector or in nearby cities. Fairbairn et al. [21] identify this "blurring of rural-urban relations through hybrid livelihoods" as one of the four most relevant themes in the field of agrarian change, while for the Mapuche specifically, Bengoa informs us that land limitations and modernization processes have compelled a constant process of migration away from rural communities, leaving behind a "remnant population" ([22], p. 75). So we encounter another set of barriers to achieving a more sustainable farm system: how far away is the labor opportunity, how often must one journey off the farm to access it, and how much does it pay? [20].

If access to agricultural markets and access to labor markets are two critical factors in peasant livelihood sustainability, they are joined by a third and arguably the most critical of all: access to land. Fairbairn et al. [21] name land and resource dispossession as perhaps the most prominent commonality across four decades of agrarian scholarship. Similarly, Amin's [23] recent theoretical dissection of food sovereignty and the peasantry zeroes in on land, land tenure, and land access as "central to debates on the future of agricultural and food production [and] peasant societies".

The very term "smallholder" hints at one component of the land problem. The World Bank defines the term as someone farming less than 2 ha of land, an amount that the International Fund for Agricultural Development labels as "often too small to ensure the nutritional well-being of the household" – and especially so if we consider the need to produce protein for a family ([24], p. 26). In addition to size, due to their socioeconomic status and their exclusion from mainstream processes of development, peasants are often "forced onto unproductive land … located in difficult environments for agriculture (arid, steep hills, etc.), often with poor soils and without access to irrigation" ([25], p. 25). This describes with eerie precision the village of Llaguepulli, with its steep hillsides, thin soils, and a virtually nonexistent irrigation infrastructure.

A final barrier worthy of note is the state, which has the power to "create favorable *or adverse conditions* for … the productive capacity of an agroecosystem" ([13], p. 53; emphasis added). A centralized government holds power over financial resources (availability of subsidies or credit), infrastructure (roads, bridges, market infrastructure), and regulations (land titling, vendor licensing, prerequisites for obtaining loans) that are crucial linchpins in a community's ability to create ecological farming systems. This has certainly been the case in Chile, where Mapuche smallholders have been actively marginalized by a state apparatus that viewed them as impediments to progress [2], and in Llaguepulli specifically, where good credit, paved roads, and clear land title have all been obfuscated or delayed according to many of my respondents.

This list of barriers is brief and far from exhaustive. I have highlighted several of the key constraints that will prove to resonate with the situation among the smallholders of Llaguepulli, in order to bring us back to the field of agroecology. All of the barriers mentioned – access to markets, to wage opportunities, to infrastructure, to credit, and, most of all, to land – can be classified as access to *resources*, and "if farmers cannot access the resources they need, … they cannot continue to maintain or develop sustainable agroecosystems." An agroecologist, therefore, must "move beyond the farm-scale to consider the … complex challenges, both social and ecological, that smallholders face in the transition towards sustainability" ([8], p. 12).

3. Research question

My fieldwork takes up this challenge by posing the following question: **What are the ecological and social conditions that motivate and that constrain the move towards a more sustainable agroecological system in a community of indigenous smallholders in southern Chile?**

4. Research site and methodology

Llaguepulli is a village consisting of approximately 40 households[1] spread out across a series of hills and valleys at latitude 38.9 degrees south, along several small bays of the undulating coastline of Lake Budi, a salty lake measuring approximately 120 km^2 in area [26]. It is part of the commune of Teodoro Schmidt in La Araucanía, the 9th region of Chile.

My introduction to the community occurred over several stages. First I established contact with a community leader during a visit in 2013, then with two employees of MAPLE Microdevelopment, a NGO that has been working closely with Llaguepulli on community-based development projects since 2012 [5]. In 2014, at a community-wide meeting, a letter I had written explaining my research intentions was presented to the community, which resulted in a written invitation from the community leadership to come and conduct a study of farming livelihoods and sustainability. My family and I arrived in November 2015 and lived there for 3.5 months, through the middle of February 2016.

I began by collecting a list of names and phone numbers of interested individuals at the first community meeting after our arrival, and after meeting with these individuals my sample grew through the "snowball sampling" technique, which consists of asking interviewees for the names of other potential respondents. Each family visit consisted of a walk around their property, followed by a semi-formal recorded interview. In total I conducted interviews with 38 individuals representing 25 households in the community (63% of households). This sample consisted of 21 adult males and 17 adult females, with an age range of 26–76 and an average age of 52. Of the 25 households, one does not engage in any agriculture – neither farming nor gardening – while the other 24 either garden or cultivate a larger acreage, or both (see **Table 1** below). Visits lasted on average 90 minutes, including the interviews which took approximately 1 h each.

In addition to interviews, I engaged in participant observation throughout the 3.5 months, which consisted of many distinct acts: attending monthly community meetings, at which the key issues facing the community are discussed at length; working with various individuals, including my two collaborators from MAPLE, to establish several community garden plots and give workshops on ecological gardening practices (see [27]); volunteering to help families plant and harvest crops, weed gardens, cut firewood, etc.; and attending semi-monthly meetings of a community-based banking group initiated by MAPLE, where community savings accounts and a microloan program are discussed. This immersion in a community's daily life exposes one to its rhythms and brings with it a distinct set of observations and insights. Before taking on the research question itself, I will present a descriptive snapshot of the community focusing on its agricultural resources and practices.

[1]The figure is imprecise due to some inhabitants splitting their time between the village and nearby cities for work, and the inherent seasonal fluctuation of the population.

	Yes	No	Other	Total
Plow	12	12	1[p]	25
Oxen	13	12	0	25
Livestock	22	3	0	25
Field Crops	20	5	0	25
Home Garden	22	3	0	25
Greenhouse	16	8	1[g]	25

[p] An elderly couple owns a plow but no longer cultivates beyond their garden.
[g] An elderly respondent has the structure of a greenhouse plastic is so shredded from age that it is nonfunctional.

Table 1. Agricultural resources owned by household (n = 25).

5. Description of the community

5.1. Landholdings

Llaguepulli is truly a community of smallholders. The average landholding per household is 2.2 ha, with a range of <0.5–7 ha, and 11 of the 25 households (44%) holding 1 ha or less.

5.2. Resources

Table 1 displays ownership of key agricultural resources by household, revealing several patterns of note. Every household except three (88%) features a home garden in which a large proportion of the family's vegetables are grown, along with a variety of flowers and herbs. Similarly, every household except three (88%) has some livestock (other than oxen for draft power), ranging from poultry to pigs to sheep, and a few households with meat cows. No households have dairy animals. A slightly lower number of households (80%) have enough arable land to cultivate field crops – most with a plow but some, on the smallest acreages, with hand tools. Just under two-thirds of households (64%) have plastic greenhouses (built by hand with locally cut wood). In terms of more basic equipment, just under half of households (48%) possess an iron plow and just over half (52%) possess a pair of oxen. The latter is a notable fact, as land is the agricultural resource in shortest supply, and maintaining a pair of oxen requires considerable space for pasture and hay. The households who do not own a plow or a set of oxen either do not have enough arable land for field crops, or they work the soil by hand, or they pay someone from outside the community to plow with a tractor.

5.3. Agricultural practices

I turn now to specific agricultural practices, and in particular those that have some bearing on sustainability. First there is the issue of growing organically. As I will discuss below, many community members mention a strong desire to free themselves from the use of chemical fertilizers and pesticides, but their ability to do so is split sharply by the scale of their cultivation. **Table 2** displays the breakdown of organic practices at two scales: the field scale (staple crops, notably potatoes and wheat and, to a lesser degree, oats and peas) and the garden scale. Of the 22 households who manage a home garden, 18 use strictly organic practices (animal manure as fertilizer, manual weed cultivation, and in some cases spreading ashes to combat blight in potatoes). At the field scale, the story is entirely the opposite – only a single family grows its field crops (potatoes, beans, and quinoa) using only composted manure, which they purchase from another community member. 17 of the 20 respondents with field crops (85%) use both conventional fertilizer and chemical pesticides for managing potato blight, while two have recently been experimenting with mixing conventional fertilizer and manure together, with an eye towards becoming all organic. The reasons for this discrepancy will be made clear later in the paper.

Aside from organic, my interviews revealed an impressively long list of other innovative techniques, listed in **Table 3**. One might assume that the list is accounted for by a small handful of innovative or entrepreneurial households, but this is not the case. Of the 25 households surveyed, fully 20 of them (80%) are represented on the list. To give another metric, the six techniques at the bottom of the list, practiced by just one household each, are spread across five different households. So there seems to be a high level of agricultural innovation oriented towards sustainability in the community.

5.4. Food self-sufficiency

There are five crops grown at scales larger than home gardens: potatoes and wheat dominate, oats are third, and peas and fava beans are a distant fourth and fifth, in terms of total acreage planted. Oats are almost exclusively grown as animal feed, while the other four are grown for human consumption (with small quantities of wheat fed to animals) or to sell; but the distinction between household consumption and market sales is critical. Almost all families make bread every day or two, and almost all of this wheat comes from their own fields. The same is true of potatoes, which are both the dominant commercial crop and a staple starch in the community's diet.

Table 4 reports the breakdown between consumption and sales for these five crops. Of the 20 households who grow potatoes and the 11 households who grow wheat, only a single

	All Organic	Hybrid*	Not Organic
Field crops organic? (n=20)	1	2	17
Home garden organic? (n=22)	18	3	1

* A mix of chemical fertilizer and organic manure

Table 2. Crop and garden management by household.

Compost Pile	14
Grow Quinoa	8*
"Manure Infusion" as Foliar Spray	5
Vermicompost Pile	4
Grow Native Potato Varieties	4
Grow Other Alternative Crops**	3
Raised Beds in Garden	2
Terraced Vegetable Beds	2
Raise Heritage Chicken Breed	1
Living Fence (trees)	1
Sprinkle Ash as Pest Deterrent	1
Use Rock Mineral Powder	1
Plant-Based Pesticides	1
Ventilation System in Greenhouse	1

* 3 of these 8 grew quinoa in the recent past but no longer did at the time of our interview.
** Includes lentils, chickpeas, jerusalem artichoke, ginger, salad greens.

Table 3. Agricultural innovations by number of household (n = 25).

household for each (the same household) reported growing the crop exclusively in order to sell it. Most of the others grow for some combination of self-consumption and market sale, and nearly half of the households report *only* growing the crop in order to consume it. For oats, peas, and fava beans the numbers are even more striking: no households grow them exclusively to sell, and over half grow them exclusively for consumption.

This striking level of self-consumption underscores the ongoing existence of the "moral economy of the peasant" [28]: the prioritization of household food security, even at the expense of turning down off-farm work that could produce more income (see [29] for a recent discussion). This in a sense provides empirical backing that the inhabitants of Llaguepulli are not merely rural residents but are peasants in the more conceptually important sense. One of my respondents articulated this relationship elegantly: "Everything that I harvest, the first priority is leaving enough for the rest of the year. Wheat, for bread. And potatoes too, I also store them to eat them. And when I need a little money, I sell something else – potatoes, or a little animal … But the first priority is leaving enough for the year." Another linked this not just to

	Potatoes	Wheat	Oats	Peas	Fava
Grow for consumption only	7	5	5	8	7
Grow primarily for consumption, sell only surplus	8	4	3	5	2
Grow for both consumption & sale	4	1	1	2	1
Grow for sale only	1	1	0	0	0
Total interviewees growing	**20**	**11**	**9**	**15**	**10**

Table 4. Field crops grown for consumption versus market sale (n = 20).

being a *campesino*, but an indigenous one: "A Mapuche wants to have a little bit of everything. For example, sheep for wool. Pigs for their meat. And other animals from time to time – for example cows to sell."

With this agricultural portrait of the community providing some background context, I turn now to the empirical research question. I break the following section into two parts, corresponding to the two halves of the query: **What are the conditions that (a) motivate and (b) constrain the move towards a sustainable agroecological system in the indigenous smallholder community of Llaguepulli, Chile?**

6. Results

Motivating factors are primarily two-fold: a deep frustration with the constant cycle of debt incurred by commodity crop production, and a strong attraction to the idea of organic farming. Constraining factors are more complex. They begin with the loss of traditional agroecological knowledge, then metastasize in the form of a lack of manure for use as organic fertilizer and a lack of animals for producing it. This insufficiency of animals stems from severe land constraints, which links all the way back to historical land dispossession by the Chilean state. In this section I will expand on each of these factors in turn.

6.1. Motivating factors

My interviews revealed one major "push factor" (an element of the current system that respondents were dissatisfied with) and one major "pull factor" (an element of an alternative system that they were drawn to) that served as motivators for seeking an agroecological alternative.

6.1.1. Push factor: cycles of debt

From an economic point of view there are two basic aspects of any farming system – costs and revenues – and on this point the agrarian studies literature is unequivocal: peasants are damned at every turn. They are often located far from primary markets, making the costs of purchasing inputs and of transporting their products higher. They are small in scale and cannot produce in volume to take advantage of the meager profit margins for commodity crops. They are economically marginalized and rarely have surplus capital or savings, making them dependent on debt for financing a given year's production. And compounding all of the above factors, they are often at the mercy of a single intermediary for supplying their inputs or purchasing their crops, or a single lender able to charge usurious interest rates [23]. Let us use the responses of community members to consider each of these three elements: costs, revenues, and indebtedness.

My very first interviewee put the matter most bluntly: "We are not happy with farming because there is no profit". I asked him and his wife if they have noticed any rise in prices, and they responded:

> *Husband: For potatoes, no. But for fertilizers, yes they go up a lot.*
>
> *Wife: Fertilizers increase a lot.*
>
> *Husband: A farmer is left in debt if he sows a lot of potatoes. Because potatoes demand a lot. You have to give it a lot of fertilizer. Then in order to control disease you have to buy all kinds of remedies. So in the end, it doesn't pay to sow a lot. And then after, you have to find a worker to help you harvest the potato. So, a lot of costs.*

Confirmations of this trend were universal among those who spoke to the cost of inputs. One farmer related that roughly 10 years before, a 50 kg sack of potato fertilizer cost 5000–6000 pesos, but by 2016 the cost had more than tripled to 20,000+ pesos (USD $27+ at the time of the research). Other respondents' recollections matched this, for example one quoting a price of 20,000 and another 22,000 for a bag of fertilizer. And this is not to even mention pesticides, which have become nearly required for growing potatoes (due to the entry of late blight into the community) and which, according to one man, "are almost more than the original chemicals for fertilizing."

Meanwhile, the costs received by the farmers are stagnant. One reason is a simple fact of economic geography: there are more fertile, flat, and well-irrigated fields elsewhere in the country where crops such as potatoes can be produced at much higher volumes and much lower cost, forcing these smallholders to lower their prices in order to compete. A second reason is more a fact of agronomy: the vast majority of potato growers in Llaguepulli are cultivating the same variety and planting it at the same time, which means it is ready for harvest at the same time, which means supply surges and the price plummets at harvest.

The main harvest window for the standard variety of potatoes stretches from mid December to late January. Earlier, in the middle of November, one woman told me that potato prices were high because most varieties had not matured yet, but she predicted that by mid-December the prices would drop "and they would not come back up again". Her husband informed me that new potatoes were selling for 30,000 pesos per 50 kg sack. Another grower told me that the first potatoes to come out of the ground (the "new potatoes" or *papas nuevas*) had been bringing 40,000 pesos per sack. At a community meeting in early December someone noted that just recently the price had dropped from 30,000 down to 22,000. On December 8 I encountered a farmer who had just sold a load of potatoes to an intermediary for the price of 15,000 per sack. On January 13 another campesino informed me the price had dropped to 8000, and by the end of January several people told me that potatoes were now selling for 5000 pesos a sack. No one I interviewed keeps careful accounting records, so it is impossible to say with certainty, but my informal calculations indicate that at a price received of 5000 pesos, a farmer is losing money on every sack sold.

And so, with farmers caught between the constant squeeze of rising input costs and stagnant farm gate prices, they are forced into a system of debt in order to fund the annual production cycle. There are only two sources for financing: a local bank, if they will lend to a Mapuche campesino, or INDAP (Institute for Farm and Livestock Development), an agency under the government's Ministry of Agriculture dedicated to the support of small-scale farmers and ranchers. Both come in for equal scorn from the inhabitants of Llaguepulli.

One interviewee said bluntly, "The banks bleed the peasant farmers" (*Los bancos desangran a los campesinos*), describing how a loan for 1 million pesos must be paid back at 30% interest, and all at once rather than month-by-month. Few farmers have this much net revenue all at once, the interviewee said, so they must request an extension, in which case the 1.3 million peso debt suddenly balloons to 2 million pesos. A 30% interest rate turns into a 100% rate. Though this narrative seems hard to believe, it was echoed by others, who noted that the moment one defaults on a payment, the interest rate skyrockets.

According to another respondent, who has actually served as an extension agent for farmers in several local communities, INDAP may not charge such blatantly usurious interest rates, but the end result is effectively the same, because after a farmer sells their crop, "the profit margin they have, they have to give it to INDAP. So, generally, they continue on in this circle. And it's too difficult to innovate or change." Much of the time an elderly peasant spent talking to me was consumed with discussing the many years during which he had labored under a debt burden to INDAP.

I would be remiss if I painted the debt problem as strictly a structural one: as with any complex issue, there is an interplay between structure and human action. One respondent, a younger woman who lives in the community part-time and has witnessed her parents' struggle with debt, expressed frustration with the way in which the agronomic conservatism of the peasant contributes:

> *It's illogical, because the inputs for growing potatoes – for example, the variety most used around here, the seeds are super expensive! And I think that no one – not even my parents, who are truly potato producers – gives concern to the costs or to whether they ultimately get any profit. I don't think they make that calculation. How much they spend, like, I don't know, 3 million [pesos] to grow the crop, and then afterwards they get 3 and a half million, it's hardly a profit. But then along comes March, and they do the whole thing over again.* (emphasis added)

What is notable is the way in which the smallholder feels yoked into the system, lacking the resources (capital, education, mobility) to leave or envision growing anything different. One pair of commentators, a widowed female landowner and her younger male farmworker, bitingly captured this sense of entrapment:

> *Respondent 1: We have to pay the debt we have. At times there isn't even enough to pay oneself. Because if the harvest is bad, and if the price is low, there isn't enough to pay oneself. And you're left short.*
>
> *Respondent 2: You have to buy the input to sell [the crop] just to pay the loan off. You're left with very little for your [household] costs. That's the reason the campesino goes no further. We can't get ahead … That's the reason the campesino is indebted. The campesino can't get ahead in any way, because the costs are large.*

6.1.2. The pull factor of organic

The "push factor" of debt is made more powerful when paired with some notion of an alternative, and my interviews revealed a strong draw towards the idea of not just sustainable agriculture generally, but organic farming in particular. It begins with the recognition that there is another way possible. Some stated it as simply as that – "We know there is an alternative – growing organically", said a middle-aged female respondent. Others framed the desire specifically as a yearning to return to an earlier form of farming, the kind practiced by their ancestors. Here are a sampling of responses from three different interviews, with the key words italicized for emphasis:

> *It would be sensational to be able to return to [organic methods] – that would be ideal. Then we would be eating more natural food.*

> *We have given a lot of thought, gone round and round, on how we can reclaim, on a small scale, our organic agriculture. With potatoes for example, we want to grow them for our own consumption, to go back to having healthy potatoes [grown] with nothing but organic compost.*

> *I feel that the land has become so tired, is so miserable, is sick, so if we get back to farming with products that are natural, with compost, like they farmed in the old days, I think that the land can be a good. It's like giving it medicine.*

Much of this seems tied to remembering farming in the "old days", as numerous respondents told stories of how their parents or grandparents cultivated plants using only animal manure as a fertilizer, or recalled how good the harvests were or how healthy the plants were before the arrival of agrochemicals. Members of 14 households specifically made such comments, even though I never asked a specific question about how their parents farmed. One respondent, probably the most well-traveled and well-educated in the village, made a link not just to his ancestors but to indigeneity:

> *It's very important because I think it can rescue the traditional ways of organic, agrarian production that Mapuche families practiced before the arrival of Occidental peoples to the South American continent – that they had been practicing since forever … Before, they didn't use pesticides, fungicides, not even chemical fertilizers.*

But respondents did not have to reminisce about a past – real or imaginary – in order to conjure the appeal of going organic. In line with the data above (see **Tables 2** and **3**), most of them have experience raising an organic vegetable garden, and some of them with growing organic crops at scale. And here the response is unequivocal. Many spoke of the superiority (in terms of both production and quality) of organic methods in the home garden. One man noted that whenever they have extra animal manure left over after fertilizing the garden they use it on the field potatoes:

> *And they give good potatoes! They're the best potatoes you can harvest. Because they live healthily. They don't come with chemicals. Super healthy. They're delicious to eat too. The difference is huge, the difference between chemicals and animal manure.*

The benefits extend beyond mere production and taste. An elderly woman related how last year she and her husband sowed beans with nothing but manure as fertilizer, and even though there was a terrible drought, they produced a "great harvest" – no doubt due to the enhanced water holding capacity that higher organic matter levels bring. And others display an insightful understanding of the mounting negative effects of agrochemical use, including the persistence of certain compounds in the soil, the unintended killing of beneficial insects and worms, and what is referred to in the sociology of agriculture as the "pesticide treadmill". One farmer, one of the most innovative in terms of experimenting with alternative crops such as lentils, noticed that late blight in his potatoes only seemed to be getting worse, rather than better, with the increased use of fungicides, and referred to it as a "cycle" that the peasant becomes trapped in.

A final quote serves to bring both the push and the pull factor together. Note how this elderly female campesina neatly ties together the stress from high input costs with the growing desire to farm organically:

> *If you could farm organically, experiment more with organic, it would be better. It would be more peaceful. You wouldn't be thinking, I have to buy who-knows-how-many sacks of fertilizer, and how much is it going to cost me. You could leave those thoughts to the side. Instead, to work with your own manure and your own thoughts, that would make you more calm, more at peace.*

All told, if we consider the number of individuals who commented on the disadvantages of farming with agrochemicals or the desire to farm organically, 30 out of 38 interviewees (79%), representing 22 of the 25 households in the study (88%), expressed this broad sentiment. This, then, brings up the obvious question that will occupy the remainder of the paper: what is holding them back? If agricultural innovations are already being practiced by a large majority of households; if there is growing dissatisfaction with the debt loads and chemical use of conventional agriculture; if there is a strong and growing desire to farm organically – what impedes Llaguepulli's smallholders from realizing this goal? The answers are neither as simple nor as obvious as they may at first appear, though they do boil down to the most basic farming resource of all: land.

6.2. Constraining conditions

6.2.1. The entrance of agrochemicals and the loss of organic farming knowledge

We might start by asking the converse question: why did the campesinos of Llaguepulli switch *away from* organic practices with the arrival of agrochemicals in the 1970s and 1980s? It is worth quoting one respondent at length, because in answering this question it he also sets the stage for later discussions:

> *Ever since the Green Revolution, people began applying chemical fertilizer. Before, it was all organic. I remember when I was a boy, 8 or 10 years old, we sowed potatoes with nothing but animal manure… Then came the famous Green Revolution, and from that point the people who had a little more money bought chemical fertilizer and applied it. And they saw that the results were fantastic. And when they saw this, they said 'Wow' and they bought more… In time there were special mixes of fertilizer for potatoes, for wheat, for oats…*

Older folks who used to collect, pile, and spread manure with their parents recalled the tremendous amount of physical labor to "make your own organic fertilizer", and when fertilizers in a bag came along, as an elderly woman said, "People became so enthused with it, and soon everyone was working with chemicals." Another effectively finished her thought: "And that's how it went, little by little, we entered that area [of agrochemical dependency]". So complete was the fertilizer transition that multiple respondents commented that the crops and even the soil itself had become "accustomed to" (*acostumbrado*) the chemicals. The oldest man I interviewed, who had been one of the earliest adopters of agrochemicals (and, it should be said, of borrowing on credit in order to purchase those chemicals), told me in plain terms, "If you don't apply fertilizer, you won't harvest anything."

The move away from "organic-by-default" farming in the span of less than a generation caused many of these techniques to fade from the collective knowledge base. One man related how, "In the old days, the old guys had corrals, where they kept all of their animals. And in these enclosures they gathered all of the manure, made big piles of it – they dedicated themselves to this… The old guys were intelligent!" But now, he went on, "the animals are just out in the fields. And the manure is lost." Speaking of composting more generally, another young farmer related, "Me personally I don't know how to make a compost pile. We're kind of at the most basic level. I imagine that there is a whole process involved, but I, at least, do not know how to make compost." A woman in her 70s told me when I first met her that she needed help with composting, and later when the discussion turned to the proper treatment and use of her own sheep pellets, said with resignation, "I don't understand much about any of that".

This lament over the loss of traditional forms of knowledge was not limited to talk of manure and composting. One woman discussed her general ignorance about soil: "We don't have the clarity – how to find the balance, what the land needs. And since we don't do soil tests, we know even less! We are working blindly." A middle-aged man related:

> *In regards to natural insecticides and pesticides, I don't have an expertise about which products I can use to control insects. I know that they exist, I know that it's possible to make a mixture of plants, herbs, and a few other products, but I don't know what the ingredients are, and what quantities to use.*

Another, younger man, when discussing the irony of having insecticides kill not only the target insect but also beneficial insects that may have eaten the pests on their own, noted, "There are so many things that one isn't able to recognize what one is doing. You think you're doing good for your soil, but actually you're doing bad."

6.2.2. Lack of animals and lack of manure

One of the questions I asked each interviewee was what they thought most impeded the community from reaching its agricultural sustainability goals. In my very first interview a peculiar theme began, as an elderly man said simply, "There aren't [enough] animals. You have to have sheep, cattle, in order to have organic manure. They used to have animals. Sheep, cows, all kept in a corral. And you gathered together all the manure. But now there aren't animals". Then, just two interviews later the same theme was echoed: the community was impeded by a lack of animals and, consequently, a lack of animal manure. Then a third interviewee reported that the thing he would *most like to do* on his farm, given the resources, would be

> *to go back to how it was before. From the animals I have, from them to get manure. And then with that same manure – because from chickens comes such a good manure! If you have poor land, you spread that manure on it … To store up manure from chickens, from geese, from ducks, from hogs – you put all that together and you make fertilizer.*

Over and over the theme returned: All told, the specific subject of (a lack of) organic fertility due to (a lack of) animal manure came up in 15 of the 25 interviews (60%).

So what accounts for the lack of animals? There seems to be a complex set of immediate reasons. Certainly one is the reason mentioned above: a fading familiarity with and interest in rearing animals and handling manure, especially on the part of the youngest generation of farmers. Members of the older generation frequently recounted how their parents or grandparents enclosed their animals in corrals each night, and over the course of a year collected large amounts of manure. Said one elderly respondent:

> *When agricultural chemicals had not arrived yet, all of our parents had corrals, where they kept the ox, some cows, and the sheep. The sheep, along with the cow's manure, makes a super-strong compost, super.*

Said another:

> *We collected it from the animal corrals where we enclosed the cows, the sheep, and the oxen, and when harvest time came, we put it in a cart and applied it… It was a lot of work but that was how we worked, because there was no other way of fertilizing.*

But this work came at a physical and time cost, as a third man noted:

> *You have to pile the manure up in the corral, different piles in the different corners, then let it sit for awhile until it's ready to use, then cultivate the land, then put all the manure in a cart, and then leave it out in the fields in little piles, and then when we went to sow the crop go along and pair the manure with the seeds. And it tired you out.*

As one older man put it, "People don't want to get their hands dirty. The work [of handling organic manure] is super dirty, they say." An older woman echoed these words almost exactly: "Now no one uses manure. They find that it's too much of a pain, it's too heavy to spread, it gets your hands dirty." And so, as a farmhand put it, "What has happened with the campesino, he has pursued comfort – regarding the question of fertilizing with fertilizer, he buys. It's easier."

The causal chain then gets more complex. With a steady loss of animal rearing and the entry of chemical fertilizers, manure becomes less valuable; the basic infrastructure of keeping animals (fences, pens) deteriorates, and in time the few animals left in the community simply stay out in the pasture. Here are three respondents from three separate interviews, all sketching in elements of this story:

> *After chemicals came along, the people did not concern themselves with keeping the animals in at night. There was no longer the necessity.*

> *We have lost the manure, we have lost that custom of closing in our animals at night. And these days, since there are hardly any anyway, we just leave them like that, free range.*

> *The animals now, day and night they are out in the pasture, so we don't gather any of that manure. But if we could close them in, for example right now at this hour of the day, by tomorrow morning we would gather a lot of manure.*

There is a final element we have not woven in yet, and that is the question of demographic pressure. When respondents referred to an older time when animals were kept and manure was collected, they were usually speaking of their grandparents – who were also some of the first Mapuche to settle these lands. Those men and women had large families – stories of 10 and more children were not uncommon – which necessitated subdividing the land so that each child could have a parcel. And those children themselves had multiple children, which has led to the situation of today: a severe land shortage. So little land, in fact, that it is virtually impossible to keep a herd of animals of a suitable size to produce manure sufficient for growing crops organically.

6.2.3. *Insufficient land*

The point is driven home by my interviewees. If answers to the question about what is impeding the community's path to sustainability were most often framed around the question of manure, the answers to a later question are even more telling. Towards the end of each interview I would ask, what does the community most need to help solve its main problems, and the answer was not "more animals", but rather, as one man put it, "Space. Fields. Land." (*Espacio. Campo. Tierra.*).

This man was one of 10 children. His parents had owned 10 ha – a considerable amount, enough to raise a family. But they divided their land equally among their children, meaning this man and his wife, who make their entire livelihood from the land, have a single hectare to farm. They have three children, the oldest of which, at 18 and just graduated from high school, would like to farm himself. As his father said: "When he is, say, 20, maybe he'll get married. He's going to have to go to the city, far away. But our idea is that no one has to live far from the family, but should be here, close by. But we can't make that happen. We need land."

And he is far from alone. One man of 48 related how his paternal grandparents farmed 24 ha – the entirety of a large hillside. That man had eight children and split the land equally among them, meaning that my respondent's father received 3 ha. His father then split the land among his own children, so my respondent farms less than 0.5 ha and manages to crowd together a small potato patch, a large vegetable garden, and a poultry pen onto a small but steep slice of hillside. That respondent's older cousin farms a similarly-sized tract of land at the bottom of the same hillside. He has six children, all but two of whom live in cities in the region. I asked if he wanted them to come back to the community and farm, and he replied simply: "No, there's nowhere to do that. I have just this little space and no more. That's why they went to the cities to work. Very little land."

Stories of agronomic constraints imposed by land limitations were a common theme in interviews. Recall what a staple crop the potato is – yet an older man, a campesino all his life, is hardly able to devote any of his land to growing them. "Those who have plenty of land, they sow lots of potatoes. Here, no." As we walked inside another man's hog enclosure, he remarked that he would love to have a market garden right there, and would be able to produce lots of vegetables. But then he would not be able to have the hogs. He must choose between the two. A younger woman and her husband decided to cease growing wheat, as they could no longer justify growing on land they needed to feed their oxen.

Oxen, the only source of field traction in the community, play a role in several others' stories of land constraints. An older couple, farmers their whole lives, related that they have to board their oxen on a property outside the community and buy hay, at a cost of 7000–8000 pesos per month (USD $10–11 at the time of the research) per animal. "By contrast, if we had sufficient land, we would have the animals on our own pasture. Or we would have them penned in, and we would have manure that we could gather." Another man, who does raise a small flock of sheep and a handful of hogs, cannot own oxen and thus must hire a man from a neighboring community to till his cropland: "Oxen I do not have… I don't have the land, the space, in order to grow the forage to feed them." He went on:

> *Many families in the community don't have oxen, for the same reason. I mean, you could have animals, but if you don't have hay stored up for the winter, what do they eat? They suffer from hunger, they get too skinny.*

The same man returned to the theme later in the interview and astutely linked the situation faced by campesinos in Llaguepulli to the socioeconomic status of the peasant smallholder more broadly:

> *The category we find ourselves in in this community is that of subsistence agriculture. A category that is very basic. We can't really even say that we're small farmers. A small farmer is someone who sows 10 hectares of potatoes – 15, 20 hectares, 50, 100. Those are small farmers. But us, no – with luck, one hectare… So if we want to be farmers in the real sense, it depends on land.*

I will leave it to one of the community's main leaders to make the final linkage, because he does it so eloquently, wrapping together the land limitations faced by the community, with the inability to subsist as a true land-based campesino, with – ultimately – the slow diminishment of the cultural identity of being Mapuche:

> *If there is not land there are no sheep, and if there are no sheep there is no wool, and if there is not wool there is no poncho (manta), there is no way to clothe yourself, we slowly lose ourselves culturally. As long as we don't have sufficient space to work the land, we seek out other options.*

And thus we come full circle, back to the theme introduced at the beginning of the paper and a hard historical truth. The Mapuche of Llaguepulli, just as virtually all Mapuche in Chile, do not own sufficient land for keeping herds of animals because they were dispossessed of their land over 100 years ago by a wave of homesteaders flocking to the Araucanía region, aided and abetted by the Chilean government. Indeed, the very word for the parcels of land which the Mapuche were deeded – often translated into English as "reservations" due to the parallel practice in North America – is more insightful if we translate it literally: "reductions" (*reducciones*). The historical Mapuche territory was "reduced" by 95% ([2], p. 46), initiating the century-long causal chain that led to the present situation.

7. Discussion and conclusion

We are tempted to look for answers to questions of agricultural sustainability in current practices and techniques and even socioeconomic circumstances, but sometimes history holds the greatest clues. A late nineteenth century traveler in Chile wrote admiringly of the *colonos* settling around the southern city of Valdivia: "They were true pioneers, who chopped down the forests, built log cabins, and planted wheat fields and orchards in their clearings. They opened up a large part of the region, and made many fine farms" ([3], p. 87). And what became of the Mapuche forcibly displaced by these settlers? The historian Florencia Mallon describes their fate, and the fate of their descendants, most incisively:

> *By reducing the amount of land and other resources controlled by the communities, the state forced the Mapuche to transform themselves from semi-migratory livestock herders into small peasant producers. The size of the original land grants … generated dramatic impoverishment by the second and third generations. Ironically, because they did not have the necessary expertise to develop intensive agriculture, Mapuche peasants ended up contributing to the degradation of their own land.* ([30], p. 238)

Apart from several of Llaguepulli's leaders, who are involved in political fights to reclaim some of their ancestral territory, no residents that I spoke to mentioned this deeper history. And yet, whether conscious of the historical roots of land dispossession or not, they made elegant links between land limitations, the cycle of agrochemical farming, and the constraints against sustainability. One offered insights into the fertilizer treadmill, how the degradation of the land forces campesinos to apply even more fertilizer than before because they are "obligated" to "squeeze the value out for their business." Another demonstrated with clarity how land limitations work directly against the adoption of organic practices:

> *I analyze the situation this way: If these days the space [that we have] is so reduced, and these days we have to provide food for our families, then we have little choice but to seek out chemical fertilizer, which is more secure. Some farmers in the community have tried to grow organically. [But] nowadays seeds are accustomed to chemical fertilizer. If you wanted to truly manage your land organically, you would need to adopt other techniques, for example you would need to rest your land. Rest your land, apply manure… And in order to do that, you need extra land at your disposition.*

But it was arguably the single poorest and most land-deprived of all my respondents who made the most impassioned cry for land. He farms approximately 0.5 ha of very steep land. He grows food only to eat it but generates no income from his land and must sell his labor outside the community to earn any revenue. When our talk turned to the issue of solving the community's most pressing needs he stated:

> *My solution would be the land. The most fundamental for me. Because it is on the land where man maintains himself. I have wanted to be a farmer since I was little, since I knew anything … I love farming, I love planting … You have to work, have to work the land. It is the mother of all of us, the land – yes! The mother of all of us! Because it is what feeds us.*

Two final caveats are in order. First, the historic deterritorialization experienced by the Mapuche has not gone uncontested. There has been a concerted effort in recent decades on the part of Mapuche communities – sometimes edging into violence – to reclaim the lands that they consider stolen [31]. And the Chilean state has recognized its historical role in Mapuche displacement and enacted reforms. Article 20 of the country's Indigenous Law established a Fund for Indigenous Land and Waters, which subsidizes the purchase of land for individuals or communities "when the area of land [currently held] by such communities is insufficient" [32]. This fund has resulted in the transfer of more than 187,000 ha to indigenous groups or individuals [33]. However, as the anthropologist Di Giminiani [34] points out, this process circumvents the ethical and political basis of Mapuche land claims, rendering them in strictly legalistic and topographical terms. More to the point, given the complexities of buying land on the private market and the limited funds available, only a small minority of applicants have actually benefitted from the fund, as "most claimants have seen their hopes for ancestral land restoration crushed" ([34], p. 490).

Second, it is not true that the inhabitants of Llaguepulli are completely without options, even in their land-deprived state. Efforts are currently underway to create possibilities for more sustainable livelihoods, with the ultimate goal of achieving a "distinctly Mapuche agroecological transition" (Krell, personal communication). These include the resurrection of traditional forms of indigenous land management appropriate to smaller spaces, such as the cultivation of adapted varieties of quinoa, the use of locally sourced seaweeds as foliar fertilizers, and processing of alternative crops such as the native *maqui* berry – as well as incorporate unique forms of economic solidarity such as community-managed microloans to encourage and support such initiatives [5]. As the readers of this volume know, agroecological solutions are possible, indeed are necessary, in land-constrained situations.

But a true process of repeasantization – that is, opening the possibility for an entire community of smallholders to once again be able to control their own agricultural fates outside the cycles of debt and agrochemical dependence – requires sufficient access to land, and such access will not simply appear with patience, nor can it be created by altering agronomic techniques. It is a solution whose realization is by its nature political: it requires a reallocation of lands formerly taken from the Mapuche. If we wish to clearly identify the causal chain that creates agro-environmental crisis, and the actions that can ameliorate it, then Naranjo's dictum [20], p. 243, is not just a social or political statement but an agroecological one: "Equitable and comprehensive land reform that benefits peasants is imperative for food sovereignty." That is how the inhabitants of Llaguepulli, and countless Mapuche communities like it in the south of Chile, will be able to feed themselves, raise their own animals, spin their own wool – in a word, reclaim the full extent of their moral economy.

Acknowledgements

I extend a heartfelt thank you to Ron Severson for deep conversations about community development and to Alison Guzman and Ignacio Krell for helping guide and provide crucial insights into this project before, during, and after fieldwork. I would also like to thank Jorge, Emilio, and Mauricio for their unwavering support in championing the project, and all the inhabitants of Llaguepulli for welcoming me into their community and homes. *¡Chaltumay kom pu che!*

Research funding

Funding from the Henry Luce III Fund for Distinguished Scholarship at the College of Wooster helped make this research possible.

Author details

Matt J. Mariola

Address all correspondence to: mmariola@wooster.edu

College of Wooster, Wooster, Ohio, USA

References

[1] Collier S, Sater WF. A History of Chile, 1808-1994. Cambridge, UK: Cambridge University Press; 1996

[2] Crow J. The Mapuche in Modern Chile: A Cultural History. Gainseville: University Press of Florida; 2013

[3] Carpenter FG. The Tail of the Hemisphere: Chile and Argentina. New York: Doubleday; 1923

[4] Caniguan N. El Budi. In: Bengoa J, editor. Mapuche: Procesos, políticas y culturas en el Chile del Bicentenario. Santiago, Catalonia; 2012. pp. 53-74

[5] MAPLE Microdevelopment. What We Do. n.d. Available from: http://www.maplemicro-development.org/maple-chile/

[6] SCImago. Agronomy and Crop Science Rankings. Jan 3, 2018. Available from: SCImago Journal & Country Rank: http://www.scimagojr.com

[7] Wezel A, Bellon S, Doré T, Francis C, Vallod D, David C. Agroecology as a science, a movement and a practice: A review. Agronomy for Sustainable Development. 2009;**29**:1-13

[8] Méndez VE, Bacon CM, Cohen R. Agroecology as a transdisciplinary, participatory, and action-oriented approach. Agroecology and Sustainable Food Systems. 2013;**37**:3-18

[9] Hecht SB. The evolution of agroecological thought. In: Altieri MA, editor. Agroecology: The Science of Sustainable Agriculture. Boulder, CO: Westview Press; 1987. pp. 1-19

[10] Gliessman S. Agroecology: Growing the roots of resistance. Agroecology and Sustainable Food Systems. 2013;**37**:19-31

[11] Altieri MA, Toledo VM. The agroecological revolution in Latin America: Rescuing nature, ensuring food sovereignty and empowering peasants. The Journanl of Peasant Studies. 2011;**38**(3):587-612

[12] La Via Campesina. Annual Report 2016. Harare, Zimbabwe: La Via Campesina; 2016

[13] De Molina MG. Agroecology and politics. How to get sustainability? About the necessity for a political Agroecology. Agroecology and Sustainable Food Systems. 2013;**37**:45-59

[14] Martínez-Torres ME, Rosset PM. Diálogo de saberes in La Vía Campesina: Food sovereignty and agroecology. Journal of Peasant Studies. 2014:979-997

[15] Bernstein H. Farewells to the peasantry. Transformation: Critical Perspectives on Southern Africa. 2003;**52**(1):1-19

[16] Van der Ploeg J. The New Peasantries: Struggles for Autonomy and Sustainability in an Era of Empire and Globalization. London: Routledge; 2008

[17] Martínez-Torres ME, Rosset PM. La Vía Campesina: The birth and evolution of a transnational social movement. Journal of Peasant Studies. 2010;**37**(1):149-175

[18] Herrera R, Lau KC, editors. The Struggle for Food Sovereignty: Alternative Development and the Renewal of Peasant Societies Today. London: Pluto Press; 2015

[19] Bernstein H. Food sovereigny via the 'peasant way': A sceptical view. The Journal of Peasant Studies. 2014;**41**(6):1031-1063

[20] Naranjo S. Enabling food sovereignty and a prosperous future for peasants by understanding the factors that marginalise peasants and lead to poverty and hunger. Agriculture & Human Values. 2012;**29**:231-246

[21] Fairbairn M, Fox J, Isakson SR, Levien M, Peluso N, Razavi S, et al. Introduction: New directions in agrarian political economy. The Journal of Peasant Studies. 2014;**41**(5):653-666

[22] Bengoa J. La agricultura y la población Mapuche. In: Bengoa J, editor. Mapuche: Procesos, políticas y culturas en el Chile del Bicentenario. Santiago, Catalonia. 2012. pp. 75-111

[23] Amin S. Food sovereignty and the agrarian question: Constructing convergence of struggles within diversity. In: Herrera R, Lau KC, editors. The Struggle for Food Sovereignty: Alternative Development and the Renewal of Peasant Societies Today. London: Pluto Press; 2015. pp. 14-34

[24] IFAD. Rural Poverty Report 2001: The Challenge of Ending Rural Poverty. Oxford: Oxford University Press; 2001

[25] Windfuhr M, Jonsén J. Food Sovereignty: Towards Democracy in Localised Food Systems. Warwickshire, UK: ITDG Publishing; 2005

[26] Universidad de Chile. Analisis del impacto economico y social y objetivos de calidad ambiental del Lago Budi. Facultad de Ciencias – Universidad de Chile, Departamento de Ciencias Ecológicas. Santiago, Chile: Gobierno de Chile, Ministerio de Obras Públicas; 2010

[27] Mariola MJ. Sustainable agriculture outreach in a marginalized Farming Village in southern Chile: Observations from a rural sociologist. International. Journal of Agricultural Extension. 2017;**5**(2):67-71

[28] Scott JC. The Moral Economy of the Peasant: Rebellion and Subsistence in Southeast Asia. New Haven, CT: Yale University Press; 1977

[29] Isakson RS. No hay ganancia en la milpa: The agrarian question, food sovereignty, and the on-farm conservation of agrobiodiversity in the Guatemalan highlands. The Journal of Peasant Studies. 2009;**36**(4):725-759

[30] Mallon F. Courage Tastes of Blood: The Mapuche Community of Nicolás Ailío and the Chilean State, 1906-2001. Durham, NC: Duke University Press; 2005

[31] Wallmapuche (Comunidades Mapuche del Wallmapu). Comunicado Público Mapuche. 2008. Available from: https://www.barilochense.com/bariloche-social/pueblomapuche/comunidades-mapuche-tomaron-direccion-nacional-de-la-conadi-en-temuco [Accessed: Jan 25, 2018]

[32] CONADI. Fondo de Tierras y Aguas. 2018. Available from: http://www.conadi.gob.cl/fondo-de-tierras-y-aguas [Accessed: Jan 25, 2018]

[33] Rivas A. Política pública de tierras y territorios indígenas. El Mostrador. Nov 2, 2014

[34] Di Giminiani P. The becoming of ancestral land: Place and property in Mapuche land claims. American Ethnologist. 2015;**42**(3):490-503

[35] Van der Ploeg J. The peasantries of the 21st century: The commoditisation debate revisited. Journal of Peasant Studies. 2010;**37**(1):1-30

[36] Corrado A. New peasantries and alternative agro-food networks: The case of Reseau Semences Paysannes. In: Bonanno A, Baker H, Jussaume R, editors. Research in Rural Sociology and Development, Vol 16. Bradford, UK: Emerald Publishing; 2010. pp. 17-30

Section 2

Soil Health, Agroforestry Systems and Sustainability

Chapter 3

Activity and Variety of Soil Microorganisms Depending on the Diversity of the Soil Tillage System

Karolina Furtak and Anna Maria Gajda

Additional information is available at the end of the chapter

http://dx.doi.org/10.5772/intechopen.72966

Abstract

Soil is an ecosystem capable of producing the resources necessary for the development of the living organisms. Soil microorganisms (bacteria and fungi) are responsible for biomass decomposition, biogenic element circulation, which makes nutrients available to plants, biodegradation of impurities, and maintenance of soil structure. The presence of microorganisms in soil depends on their chemical composition, moisture, pH, and structure. Human activity has an indispensable influence on the formation of ecosystems. Soil tillage has an impact on the chemical and physical parameters of the soil, and thus on its biological properties. The use of inappropriate agro-technology can lead to degradation of the soil environment. Changes in soil properties may cause changes in soil abundance, activity, and diversity. Cultivation can affect microorganisms, causing their mortality and reducing the availability of nourishment in the soil. Therefore, it is extremely important to assess the diversity and microbiological activity of soil in relation to soil-tillage technology.

Keywords: agricultural, biodiversity, soil microorganisms, soil quality, tillage system

1. Introduction

The soil is the most outer layer of lithosphere. It covers almost the entire surface of the land; in it, there is a circulation of nutrients, biotic, and abiotic processes. Soil is a key element of the environment. It performs many functions in the environment (**Figure 1**). The five main ones are listed [1] below:

1. Environmental formation—shaping the local climate, relief of the earth's surface, natural reservoir of water resources retention;
2. Ecological—an element of circulation of biogenic elements and organic matter retention;

3. **Edaphic**—creation of living conditions for organisms (microorganisms, plants, animals);
4. **Protection**—counteracting changes in the environment by means of sorption properties, *e.g.*, absorbing toxic compounds;
5. **Economic**—place of work for man, a farmer.

This is a general separation of soil functions. Each of them can be discussed separately. All soil functions are affected by human farming. This chapter is an attempt to collect information on the influence of man on the edaphic function of soil.

The soil has the capacity to self-produce resources necessary for the development of living organisms [2]. This makes it a living environment for over 30% of the species existing on Earth. Soil organisms are a very important component of soils and are referred to as "*biological engine of the earth*" [3]. Soil quality can be defined as the balance between high activity and high microbiological biodiversity [4]. Soil quality plays an important role in protecting the environment, preserving biodiversity and good agricultural practices [5–7]. Agriculture has an impact on soil, and thus on the qualitative and quantitative composition of soil microorganisms.

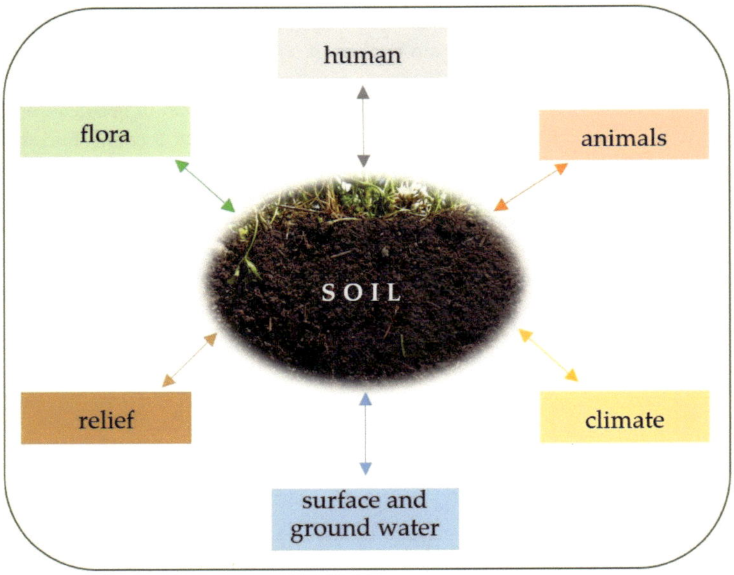

Figure 1. Soil functions in environment.

2. Soil microorganisms

Soil microorganisms are involved in many biogeochemical processes. They are a very important functional group of soil organisms. They are responsible for mineralisation of organic matter, element circulation, synthesis of proteins, and nucleic acids, as well as transformation of phosphorus forms. Rhizosphere microorganisms increase plant health and can protect against pathogens [8, 9] (**Figure 2**).

Agricultural biodiversity concerns with indigenous bacteria, plants, fungi, animals, and their equivalents introduced into the agricultural space by man and is related to the structure of soil and its use [10, 11]. It is estimated that from 1 g of soil, about 4000–6000 different bacterial genomes can be isolated [12]. Classic microbial analyses of microorganisms allow to isolate 0.1–10% of the population of bacteria present in the environment. The other species are not reared, which means that they cannot be reared under laboratory conditions. In research on the presence of microorganisms in soil, it is important not only to evaluate their quantitative and qualitative assessment, but above all to perform their functions, their role in the ecosystem, and their impact on other organisms.

The biological activity of the soil depends on the correct number and species composition of microorganisms and their enzymatic activity. Microorganisms mediate 80–90% of all processes occurring in the soil [13]. They create favorable conditions for germination of seeds and growth of the root system of cultivated plants, which is very important for a high yield [5]. Plants emit a large number of various chemical compounds into the soil, which shape the composition of microorganisms in the environment. Microorganisms use these root secretions as a source of food. The rhizosphere is a habitat mainly for bacteria and mycorrhizal fungi. Some microorganisms may produce antibiotics that block harmful microorganisms. In addition, soil microorganisms can also improve the condition of plants by releasing growth regulators (e.g., ethylene, auxin, and cytokine) and making available some nutrients (e.g., phosphorus). Polymer-producing microorganisms can improve the soil structure. Among the significant soil microorganisms, it is worth mentioning the bacteria of the genus *Pseudomonas* sp., bacteria that inhabit the root zone of plants [14]. Bacteria of this kind produce various

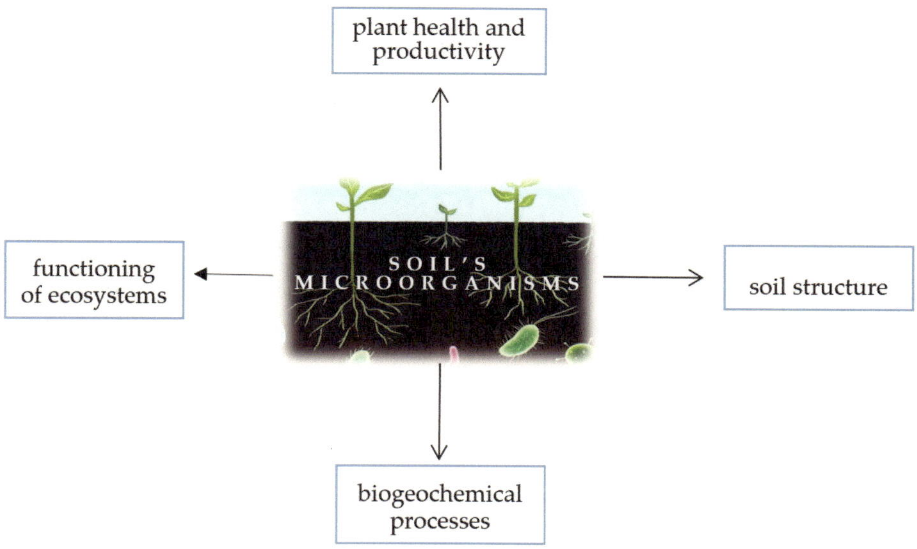

Figure 2. Functions of soil microorganisms.

biologically active compounds such as antibiotics and lytic enzymes, as well as ethylene, auxin, and gibberellin. In addition, *Pseudomonas* compete for nutrients with pathogenic microorganisms, *e.g.* for iron by creating a siderophores. The bacteria binding atmospheric nitrogen are also important for the cultivation of plants such as *Azotobacter*, *Clostridium*, *Rhizobium*, and *Bradyrhizobium* [15].

3. Impact of agricultural activity on the soil quality

Human activities in the natural environment, above all the intensification of agriculture and the use of plant protection products, change the activity and diversity of the soil environment. This can lead to changes in the functioning and sustainability of the ecosystem [9, 16]. Changes in soil quality lead to changes in plant variety and productivity [17]. Human activity is aimed at giving the soil the best possible soil production properties in order to increase the yield of plants.

Proper agricultural management should take into account the microbiological and physicochemical properties of the soil. The activity of microorganisms and their diversity is a sensitive indicator of soil quality, which is the subject of research for many researchers. The cultivation systems applied by man modify the quantitative and qualitative composition of organisms inhabiting the soil environment [18]. In [19], it was found that soil microorganisms are sensitive to anthropogenic factors, in particular agricultural activity.

Numerous studies indicate that agricultural treatments strongly change chemical [5, 20–22] and physical [23, 24] parameters of soil. Changes in soil physicochemical parameters can lead to changes in soil microbial diversity [25, 26]. Some researchers suggest that the agricultural techniques generate a homogeneous soil environment and reduce bacterial diversity [27]. Other studies [5] indicate that the soil has a high capacity to maintain biodiversity. Numerous studies are being carried out to compare the impact of soil-tillage systems on soil microbiology [8, 22, 25, 26, 28]. Some studies confirm that agricultural practices have an impact on the soil composition of microorganisms [29, 30]. Other researchers suggest that changes in the structure of the bacterial population are mainly due to chemical properties and soil structure [5]. The research [31] also reports that the soil has buffering capacity and that composition and functionality of microorganisms depends on the type of soil.

It is well known that the biodiversity of microorganisms varies from untapped soil to land cultivated in agricultural areas. In [20], it was demonstrated that in uncultivated soils, the obtained number of operational taxonomic units was about 1.5 million, 30% higher than agricultural soils. Similar results were obtained by [32] who noted that 140–150 species per 1 g of soil are found in the soil cultivated for agriculture, while in natural soils, this amount is about 1000 species per 1 g of soil. It is estimated that about 40% of the world's agricultural land has already been degraded [22, 33, 34].

3.1. The ecological farming systems

Sustainable agriculture is based on supporting natural biological processes without interfering with the environment [35]. In simplified systems, the cultivation procedures are minimised. They consist mainly in loosening of the external soil layer, leaving crop residues and the use of stubble intercrops. Leaving harvest residues has an impact on soil nutrient management, which can be beneficial for the biological properties of soil [36]. Among the organic systems, there is also the use of no-tillage, which is limited to the creation of a seed groove for sowing. Studies have shown that simplifications lead to an increase in the number of microorganisms of the genera *Cladosporium* and *Mucor*, which may inhibit the development of pathogens in soil. In addition, two to three times higher biomass of earthworms in the soil under direct sowing was recorded [37]. Tests carried out by [38] demonstrated that organic farming contributes to preserving soil biodiversity. In addition, the beneficial effect of organic fertilisation on soil microbiology is well known and documented [39, 40].

3.2. The conventional and intensive farming systems

Traditional cultivation systems are based on specialised agricultural machinery that directly interferes with the soil structure. The deep ploughing up to 35 cm deep into the soil can cause negative consequences in the physical, chemical, and biological properties of the soil [41, 42]. Numerous studies have shown that conventional soil tillage and monoculture can have a negative impact on soil quality [28, 43, 44]. Intensive ploughing leads to a more compact soil, which may result in lower yields. Intensive plant production and multiarea farms very often fail to take environmental protection into account and can have a detrimental effect on soil quality [44]. In intensive agriculture, many mineral fertilisers, antibiotics, and hormones are used, which leads to degradation of the soil environment. At the same time, it carries a risk of biodiversity loss. The intensive mixing of the soil with the spacing plough causes losses of organic matter by accelerating the humus mineralization process. Such systems aggravate the network of interactions between soil organisms and aggravate disease-related problems [25, 45]. Many years of conventional long-term soil cultivation may also contribute to the decline of microbial population in soil.

4. Methods for analysing the impact of human agricultural activities on soil quality

Research on soil quality is very important in terms of the quality of the yields obtained and the environmental impact assessment of agriculture. Soil as a habitat for many organisms and a place where many biochemical processes occur is sensitive to natural and abiotic factors, including agricultural activities. Soil quality assessment is based on an analysis of the physical-chemical and microbiological parameters of the soil (**Table 1**).

Parameter	Method	Literature
Soil microbial biomass carbon (MBC) and nitrogen (MBN) contents	The fumigation-extraction (F-E) and fumigation-incubation (F-I) method	[46]
Potentially mineralizable nitrogen (PMN)	The short-term (waterlogged) anaerobic incubation method	[47]
Dehydrogenases activity	Method with 2,3,5-triphenyltetrazolium chloride (TTC) as a substrate	[48]
Acid and alkaline phosphatases activity	Method with p-nitrophenyl phosphate (PNP) as a substrate	[49]
Urease activity	Measuring the concentration of an ammonium	[50]
3,6-diacetylfluorescein (FDA) hydrolysis activity	Measuring the concentration of fluorescein	[51]
Soil organic carbon (SOC)	Oxidation with a mixture of potassium dichromate and sulphuric acid	[52]
Strength of soil aggregates	Strength testing device	[53]
Community level physiological profiling	Biolog ECOplate (Biolog Inc., Hayward, USA)	[54]

Table 1. Selected soil quality parameters and methods of their evaluation.

Research on soil quality analysis:

- Determination of enzymatic activity, e.g., dehydrogenases, acidic and alkaline phosphatase, protease;
- Identification of soil microorganisms by molecular and immunological means through sequencing, hybridisation analysis, Elisa test, PCR-based reactions;
- Analyses of soil physical-chemical parameters e.g., pH, EC, granulometric composition, water volume;
- Evaluation of functional diversity of soil microorganisms by the CLPP method—*Community Level Physiological Profiling.*

However, despite many studies and the variety of methods used, there is still no single, universal parameter that would allow the quality of the soil to be determined quickly. In addition, it is necessary to carry out long-term studies, because only in this way can changes in the environment be noticed.

5. Impact of different farming systems on selected soil quality parameters

5.1. The enzymatic activity

Enzyme activity is considered to be one of the main parameters expressing the overall microbiological activity of the soil. The activity of dehydrogenase, peroxidase, phosphatase, protease and urease, β-glucosidase, catalase, cellulase, and invertase is commonly indicated. These enzymes are mediators and catalysts in many biochemical processes in soil, including formation and decay of humus, molecular nitrogen fixation, nitrification and denitrification, phosphorous compounds transformations, or detoxification of xenobiotics. The presence of enzymes and their activity are evidence of soil metabolism, while being sensitive to environmental changes.

Simplified systems have a high dehydrogenase activity [55–57]. Restriction of plough use has a positive effect on the activity of these enzymes [58]. In direct sowing soils, dehydrogenases also have a higher dehydrogenase activity than conventional crops [59].

Phosphatases catalyse the hydrolysis of organic phosphorus compounds once again responsible for its management in plants. They can be used as an indicator of the potential rate of phosphorus compounds mineralization. They are also very sensitive to heavy metal contamination. For acid phosphatase, higher values are recorded in soils from conventional crops and monoculture, e.g., the United States of America. Areas of the European Union's main agricultural holdings are covered by the so-called maize and alkaline phosphatase in simplified crops [36, 60–62].

The FDA hydrolysis is related to the transformation of organic matter into soil. Many studies have shown that its value is higher in soils than in simplified, organic crops [43].

Urease activity is also sensitive to the way soil tillage is cultivated. It participates in the conversion of nitrogen compounds in soil [63]. In [64], it was showed that the soils from simplified crops show up to two times higher urease activity than those from conventional crops. Similar results were obtained by [65].

5.2. The soil organic carbon content (SOC)

Soil organic carbon content is a sensitive soil quality indicator [66, 67]. The soil microbial biomass carbon (MBC) and nitrogen (MBN) represent a small fraction of the total soil organic C (1–5%) and N (2–6%) [68].

Intensive monoculture of rice causes a decrease in SOC [44]. Less intensive soil tillage improves the stability of soil aggregates and increases the SOC concentration [69]. Researchers also observe higher SOC values in soil, where ploughing is not used and plant residues are left on the surface [70]. In the case of maize, researchers have found that conventional cultivation reduces the amount of SOC in soil compared to zero tillage [71]. Similar results were obtained in the case of winter wheat cultivation, where the SOC content in the upper soil layers was up to 23% higher than in the case of conventional seedlings [28, 72]. Conventional cultivation increases SOC sequestration [71]. Research [73] has shown that in agricultural soil can contain 20% less SOC and MBC than in soil under indigenous grassland.

5.3. The microbial biomass carbon (MBC) and nitrogen (MBN) content

Determination of the soil microbiological biomass content is one of the basic parameters for monitoring soil changes. Soil management has a large impact on the size of the biomass pool of microorganisms [61, 62].

There is a negative effect of conventional crops on the carbon and nitrogen content of microbial biomass [55, 62]. Whereas, the soil under simplified and organic cultivation has a high MBC and MBN content [55, 57, 64]. The MBC and MBN in direct sowing soil are also higher than in conventional sowing [74]. In several years of research [75], the number of MBC in conventional soil cultivation decreased, with a steady level of biomass in reduced tillage and direct sowing.

5.4. The functional analysis

Soil microorganisms are complex, multispecies communities. The processes taking place with their participation should be analysed as a sum of functions of the entire population and not as individual species [54, 76]. Using the Biolog Ecoplate, the [43] observed a greater metabolic diversity in soil than in reduced tillage compared to conventional tillage. Similar results were obtained by [77, 78]. In soils from urban ecosystems, the highest microbiological activity was observed in roadside-tree soils, while the lowest activity was observed in forest sites. At the same time, the authors did not notice any significant differences among long-term maize production, park sites, and street green sites [79].

The higher average utilisation intensity (AWCD) values were recorded in soil under conventional tillage with residue retention, than zero tillage and residue removal, after 15 years of cultivation. At the same time, it has been shown that the higher functional diversity of microorganisms is in soil under wheat cultivation compared to soil under maize cultivation [74]. Research [80] showed that ability of the microbial communities to utilise substrates on ECOplate was affected by fertilisation, but not by crop rotation treatments. At the same time, it has been demonstrated that the long-term use of manure can contribute to increase the beneficial functional diversity of microorganism in agricultural soils. It was demonstrated that the functional diversity of microbial community depended on the presence of plant species [81]. The CLPP analysis also showed that the soil under cultivation of transgenic rice did not differ significantly in functional diversity of microbial communities compared to control [82].

5.5. The biodiversity in soils with different uses

The agricultural activity of mankind affects many of the physical and chemical properties of soil and its microbiological activity. Depending on the manner of soil utilisation, there are differences in the composition of the population of microorganisms inhabiting the soil. The [83] concluded that the species diversity of soil microorganisms is primarily influenced by the abiotic soil properties independent of location or land-use type. They demonstrated that location and land-use type together explained more than 33% of the variation in soil biota diversity. However, the [84] have shown that soils with low human participation rate (cork-oak, forest, and pasture) were characterised by a more stable bacterial environment than soils with a high human participation rate (vineyards and managed meadow). Research conducted by [21, 22] also shows that soils used for agricultural use are degraded, which is noted in lower microbiological parameters when compared with uncultivated soils. Changes in soil communities and loss of soil biodiversity threaten the multifunctionality and sustainability of the ecosystem [17].

Large variations in *Proteobacteria* composition are observed in soils with different uses. There were also changes in the composition of bacterial communities in the case of a change in the way forest soil are used to cultivate crops and grazed pastures [84, 85]. Studies conducted in Poland showed that in soils used for agricultural use, there was a significantly lower amount of ammonifying bacteria in comparison with soils not used for agricultural purposes [21, 22]. Similar results were obtained in other studies on agricultural soils [86, 87].

It is well known that more intensively managed soils often contain less fungi biomass [88]. Lack of ploughing while leaving plant residues on the soil surface may increase the amount of fungi pathogenic to plants (*Fusarium* sp.). At the same time, a long-term economy without ploughing leads to a deep diversification of microbial communities and also develops high fungal biomass in surface soil [89]. Other researchers have shown that cultivation practices had little impact on the diversity of microbial communities [90].

Nitrogen fertilisation of soil influences the number of individual groups of microorganisms in soil. High nitrogen doses may lead to the accumulation of toxic substances, e.g., ammonia in soil, which does not have a beneficial effect on the *Actinobacteria* [91].

Collections of microorganisms decomposing cellulose and hemicellulose change the composition of the population in response to changes in the composition of organic fertilisers [39]. The use of manure as a fertiliser in rice fields increased the activity of microorganisms, with a simultaneous decrease in this activity in the case of chemical fertilisers [92].

Higher mite density, *Collembola* and *Myriapoda*, were observed in forest ecosystems, but on grasslands more earthworms were observed [88].

The examples given above show that soil is a complex ecosystem, and the cultivation method has an impact on many soil parameters (biological, chemical, physical), and thus also the organisms that inhabit it. Changes in populations of soil organisms may have an impact on plant growth and thus on yields.

6. Conclusions

It is well known that the agricultural activity of man has an impact on the soil environment, while changes in the soil shape changes in the whole accompanying ecosystem. Agriculture is an indispensable part of the economy and production, but it is up to man to decide whether he is cultivating the soil in order to increase harvests for only 1 year or for further perspective. The condition of the soil and its quality are deteriorating over time. Many effects of human activity are visible not sooner than in a few or even several years. For this reason, it is worthwhile to carry out monitoring of agricultural land in order to observe how quickly and to what extent changes occur. It is important to manage the environment sensibly so that it does not degrade and at the same time allows to obtain good production results. An additional problem is the difficulty in interpreting and comparing results of biogeochemical parameters with new methods of microbial identification (e.g., sequencing). Therefore, the subject of changes in agricultural soil is a challenge and objective of agricultural science research.

Acknowledgements

Supported by the Task 1.3 of the Multi-Annual Programme of the Institute of Soil Science and Plant Cultivation—State Research Institute, Puławy (IUNG-PIB) and by the Statute Project Nos. 2.26 and 2.38.

Author details

Karolina Furtak* and Anna Maria Gajda

*Address all correspondence to: kfurtak@iung.pulawy.pl

Department of Agriculture Microbiology, Institute of Soil Science and Plant Cultivation—State Research Institute, Puławy, Poland

References

[1] Degórski M. Geografia gleb jako dyscyplina fizycznogeograficzna. Przegląd Geograficzny. 2004;**3**(76):273-274

[2] Russel S. Znaczenie badań enzymów w środowisku glebowym. Acta Agrophysica, Rozprawy i Monografie. 2005;**3**:5-9

[3] Haygarth PM, Ritz K. The future of soils and land use in the UK: Soil systems for the provision of land-based ecosystem services. Land Use Policy. 2009;(26):187-197

[4] Li P, Zhang T, Wang X, Yu D. Development of biological soil quality indicator system for subtropical China. Soil and Tillage Research. 2015;**126**:112-118

[5] Girvan MS, Bullimore J, Pretty JN, Osborn AM, Ball AS. Soil type is the primary determinant of the composition of the total and active bacterial communities in arable soils. Applied Environmental Microbiology. 2003;**69**:1800-7809

[6] Lemaire G, Franzluebbers A, Carvalho PC, Dedieu B. Integrated crop-livestock system: Strategies to achieve synergy between agricultural production and environmental quality. Agriculture Ecosystems and Environment. 2014;**190**:4-8

[7] Martinez-Selgado MM, Gutierrez-Romero V, Jannsens M, Ortega-Blu R. Biological soil quality indicators: A review. In: Mendez-Vilas A, editor. Current Research, Technology and Education Topics in Applied Microbiology and Microbial Biotechnology. Microbiology Series N 2. 1st ed. Badajoz: Formatex Research Center; 2010. pp. 319-328

[8] Abigail A, Salyers D, Whitt D. Ziemia: Planeta mikroorganizmów. In: Markiewicz Z, editor. Mikrobiologia. Różnorodność, chorobotwórczość, środowisko. Warszawa: PWN; 2005. pp. 3-5

[9] Nannipieri P, Ascher J, Ceccherini MT, Landi L, Pietramellara G, Renella G. Microbial diversity and soil functions. European Journal of Soil Science. 2003;**54**:655-670

[10] Arias ME, Gonzales-Perez JA, Gonzales-Vila FJ, Ball AS. Soil health-a new challenge for microbiologists and chemists. International Microbiology. 2005;**8**:13-21

[11] UNEP/CBD/SBSTTA/7/12. Developing indicators for national-level monitoring of biodiversity. Draft prepared by the expert group on indicators of biological diversity including indicators for rapid assessment of inland water ecosystems [dissertation]. Montreal; 2003

[12] Torsvik V, Torsvik VL, Sørheim R, Goksoyr J. Total bacterial diversity in soil and sediment communities: A review. Journal of Industrial Microbiology. 1996;**17**:170-178

[13] Nannipieri P, Badalucco L. Biological processes. In: Bembi DK, Nieder R, editors. Handbook of Processes in the Soil–Plant System: Modelling Concepts and Applications. Binghamton: The Haworth Press; 2003. pp. 57-76

[14] Sørensen J, Nybroe O. Pseudomonas in soil environment. In: Juan-Luis R, editor. Pseudomonas: Genomics, Life Style and Molecular Architecture. Boston: Springer; 2004. pp. 369-401. DOI: 10.1007/978-1-4419-9086-0

[15] Gałązka A, Bigos J, Siebielec S. Plant growth promotion by bacteria of the genus *Azospirillum* and their application in agriculture. Polish Journal of Agronomy. 2015;**23**:48-62

[16] Gajda AM. Microbiological and Biochemical Indices of Quality of Soils under Winter Wheat Grown in Different Tillage Systems. Puławy: Monografie i Rozprawy Naukowe, IUNG; 2015

[17] Wagg C, Bender SF, Widmer F, Van der Heijden MGA. Soil biodiversity and soil community composition determine ecosystem multifunctionality. PNAS. 2014;**111**(14): 5266-5270

[18] Kladivko EJ. Tillage systems and soil ecology. Soil and Tillage Research. 2001;**61**:61-76

[19] Kuffner M, Pinar G, Hace K, Handschur M, Haslberger AG. DGGE-fingerprinting of arable soils shows differences in microbial community structure of conventional and organic farming systems. Journal of Food Agriculture and Environment. 2004;**2**:259-267

[20] Wolińska A, Górniak D, Zielenkiewicz U, Goryluk-Salmonowicz A, Kuźniar A, Stępniewska Z, Błaszczyk M. Microbial biodiversity in arable soils is affected by agricultural practices. International Agrophysics. 2017;**31**:259-271. DOI: 10.1515/intag-2016-0040

[21] Wolińska A, Szafranek-Nakonieczna A, Banach A, Błaszczyk M, Stępniewska Z. The impact of agricultural soil usage on activity and abundance of ammonifying bacteria in selected soils from Poland. SpringerPlus. 2016;**5**:565-578. DOI: 10.1186/s40064-016-2264-8

[22] Wolińska A, Szafranek-Nakonieczna A, Zielenkiewicz U, Tomczyk-Żak K, Banach A, Błaszczyk M, Stępniewska Z. Quantified characterization of soil biological activity under crop cultivation. Journal of Advances in Biology. 2016;**8**(3):1655-1665

[23] Ozgoz E, Gunal H, Acir N, Gokmen F, Birol M, Budak M. Soil quality and spatial variability assessment of land use effects in a Typic Haplustoll. Land Degradation & Development. 2013;**24**:277-286. DOI: 10.1002/ldr.1126

[24] Singh K, Mishra AK, Singh B, Singh RP, Patra DD. Tillage effects on crop yield and physicochemical properties of sodic soils. Land Degradation & Development. 2016;**27**(2): 223-230. DOI: 10.1002/ldr.2266

[25] Kheyrodin H, Ghazvininan K, Taherian M. Tillage and manure effect on soil microbial biomass and respiration, and on enzyme activities. African Journal of Biotechnology. 2012;**11**(81):14652-14659. DOI: 10.5897/AJB09.1335

[26] Tintor B, Miloŝević N, Vasin J. Microbiological properties of chernozem of southern Backa (Serbia) according to different methods of land use. Field and Vegetable Crop Research. 2009;**46**:189-198

[27] Palmer KM, Young JPW. Higher diversity of *Rhizobium* leguminosarum biovar viciae populations in arable soils than in grass soils. Applied Environmental Microbiology. 2000;**66**:2445-2450

[28] Gajda AM, Czyż EA, Stanek-Tarkowska J, Dexter AR, Furtak KM, Grządziel J. Effects of long-term tillage practices on the quality of soil under winter wheat. Plant Soil Environment. 2017;**63**(5):236-242. DOI: 10.17221/223/2017-PSE

[29] Bossio DA, Girvan MS, Verchot L, Bullimore J, Borelli T, Albrecht A, et al. Soil microbial community response to land use change in an agricultural landscape of western Kenya. Microbial Ecology. 2005;**49**:50-62

[30] Imaz MJ, Virto I, Bescansa P, Enrique A, Fernandez-Ugalde O, Karlen DL. Soil quality indicator response to tillage and residua management on semi-arid Mediterranean cropland. Soil and Tillage Research. 2010;**107**:17-25. DOI: 10.1016/j.still.2010.02.003

[31] Grządziel J, Gałązka A. Microplot long-term experiment reveals strong soil type influence on bacteria composition and its functional diversity. Applied Soil Ecology. 2017. Article in press. http://dx.doi.org/10.1016/j.apsoil.2017.10.033

[32] Tscharntke T, Clough Y, Wanger TC, Jackson L, Motzke I, Perfecto I, et al. Global food security, biodiversity conservation and the future of agricultural. Biological Conservation. 2012;**151**:53-59. DOI: 10.1016/j.biocon.2012.01.068

[33] Shrestha K, Stevens S, Shrestha P, Adetutu EM, Walsh KB, Ball AS, Midmore DJ. Characterization of the soil microbial community of cultivated and uncultivated vertisol in Australia under several management regimes. Agriculture, Ecosystems and Environment. 2015;**199**:418-427

[34] Wolińska A, Kuźniar A, Zielenkiewicz U, Izak D, Szafranek-Nakonieczna A, Banach A, Błaszczyk M. Bacteroidetes as a sensitive biological indicator of agricultural soil usage revealed by culture-independent approach. Applied Soil Ecology. 2017;**119**:128-137

[35] Gałązka A, Łyszcz M, Abramczyk B, Furtak K, Grządziel J, Czaban J, Pikulicka A. Biodiversity of Soil Environment: Overview of Parameters and Methods in Soil Biodiversity Analyses. Puławy: Monografie i Rozprawy, IUNG; 2016. 100p

[36] Gałązka A, Gawryjołek K, Grządziel J, Frąc M, Księżak J. Microbial community diversity and the interaction of soil under maize growth in different cultivation techniques. Plant Soil Environment. 2017;**63**:264-270

[37] Lenart S, Sławiński P. Wybrane właściwości gleby oraz występowanie dżdżownic w warunkach siewu bezpośredniego i płużnej uprawy roli. Fragmenta Agronomica. 2010; **27**(4):86-93

[38] Caporali F, Mancinelli R, Campiglia E. Indicators of cropping system diversity in organic and conventional farms in central Italy. International Journal of Agricultural Sustainability. 2003;**1**(1):6772

[39] Dumontet S, Cavoski I, Ricciuti P, Mondelli D, Jarrar M, Pasquale V, Crecchio C. Metabolic and genetic patterns of soil microbial communities in response to different amendments under organic farming system. Geoderma. 2017;**296**:79-85

[40] Zhang QC, Shamsi IH, Xu DT, Wang GH, Lin XY, Jilani G, Hussain N, Chaudhryn AN. Chemical fertilizer and organic manure inputs in soil exhibit a vice versa pattern of microbial community structure. Applied Soil Ecology. 2012;**57**:1-8

[41] Busari MA, Kukal SS, Kaur A, Bhatt R, Dulazi AA. Conservation tillage impacts on soil, crop and environment. International Soil and Water Conservation Research. 2015;**3**:119-129

[42] Derpsch R, Friedrich T. Development and current status of no-till adoption in the world. In: 18th Triennial Conference of the International Soil Tillage Research Organization (ISTRO); June 15-19, 2009; Izmir. Turkey: 2009. pp. 1-13

[43] Gajda AM, Czyż EA, Dexter AR. Effects of long-term use of different farming systems on some physical, chemical and microbiological parameters of soil quality. International Agrophysics. 2016;**30**:165-172

[44] Tran Ba L, Le Van K, Van Elsacker S, Cornelis W. Effect of cropping system on physical properties of clay soil under intensive rice cultivation. Land Degradation and Development. 2016;**27**:973-982

[45] Marquad VT, Roscher C, Schumacher J, Buchmann N, et al. Plant species richness and functional composition drive over yielding in a 6-year grassland experiment. Ecology. 2009;**90**:3290-3302

[46] Jenkinson DS, Powlson DS. The effects of biocidal treatment on metabolism in soil V. A method for measuring soil biomass. Soil Biology and Biochemistry. 1976;**8**:209-213

[47] Waring SA, Bremner JM. Ammonium production in soil under waterlogged conditions as an index of nitrogen availability. Nature. 1964;**201**:951-952

[48] Casida LE, Klein DA, Santoro T. Soil dehydrogenase activity. Soil Science. 1964;**98**:371-376

[49] Tabatabai MA, Bremner JM. Use of p-nitrophenyl phosphate for assay of soil phosphatase activity. Soil Biology and Biochemistry. 1969;**1**:301-307

[50] Tabatabai MA, Bremner JM. Assay of urease activity in soils. Soil Biology and Biochemistry. 1972;**4**:479-587

[51] Dick RP, Breakwell DP, Turco RF. Soil enzyme activities and biodiversity measurements and integrative microbial indicators. In: Doran JW, Jones AJ, editors. Methods of Assessing Soil Quality. Madison: Soil Society of America Publication; 1996. pp. 247-272

[52] Walkley A, Black IA. An examination of the degjareff method for determining SOM and a proposed modification of the chromic acid titration method. Soil Science. 1934;**37**:29-38

[53] Dexter AR, Kroesbergen B. Methodology for the determination of tensile strength of soil aggregates. Journal of Agricultural Engineering Research. 1985;**31**:139-147

[54] Garland J, Millis A. Classification and characterization of heterotrophic microbial communities on the basis of patterns of community-level sole-carbon-source utilization. Applied Environmental Microbiology. 1991;**57**:2351-2359

[55] Furtak K, Gawryjołek K, Gajda AM, Gałązka A. Effects of maize and winter wheat grown under different cultivation techniques on biological activity of soil. Plant Soil Environment. 2017;**63**(10):449-454. DOI: 10.17221/486/2017-PSE

[56] Furtak K, Gajda AM. Activity of dehydrogenases as an indicator of soil environment quality. Polish Journal of Soil Science. 2017;**50**(1):33-40. DOI: 10.17951/pjss/2017.50.1.33

[57] Sikora S, Mrkonjic M, Kisic J. Importance of the soil microbial state-experience from the south-east European region. In: Miransari M, editor. Soil Tillage and Microbial Activities. India: Research Signpost; 2011. pp. 145-154

[58] Marinari S, Mancinelli R, Campiglia E, Grego S. Chemical and biological indicators of soil quality in organic and conventional farming systems in Italy. Ecological Indicators. 2006;**6**:701-711

[59] Majchrzak L, Niewiadomska A, Natywa M. Evaluation of dehydrogenase activity of spring barley depending on the tillage system, previous crop and type of crop residua. Annales Universitatis Mariae Curie-Skłodowska Lublin-Polonia. 2014;**69**(4):103-111

[60] Gałązka A, Gawryjołek K, Perzyński A, Gałązka R, Księżak J. Changes in enzymatic activities and microbial communities in soil under long-term maize monoculture and crop rotation. Polish Journal of Environmental Studies. 2017;**26**(1):39-46

[61] Masto RE, Chhonkar PK, Singh D, Patra AK. Changes in soil biological and biochemical characteristics in long-term field trial on a sub-tropical inceptisol. Soil Biology and Biochemistry. 2006;**38**:1577-1582

[62] Winding A, Hund-Rinke K, Rutgers M. The use of microorganisms in ecological soil classification and assessment concepts. Ecotoxicology and Environmental Safety. 2005;**62**:230-248

[63] Izquiergo I, Caravaca F, Alguacil M, Hernandez G, Roldan A. Use of microbiological indicators for evaluating success in soil restoration after vegetation of mining area under subtropical conditions. Applied Soil Ecology. 2005;**30**:3-10

[64] Kabiri V, Raiesi F, Ghazavi MA. Tillage effects on soil microbial biomass, SOM mineralization and enzyme activity in a semi-arid Calcixerepts. Agriculture, Ecosystems and Environment. 2016;**232**:73-84

[65] Mohammadi K, Heidari G, Javaheri M, Karimi-Nezhad MT. Soil microbial response to tillage systems and fertilization in a sunflower rhizosphere. Archives of Agronomy and Soil Science. 2013;**59**:899-910

[66] Tesfay A, Cornelis WM, Nyssen J, Govaert B, Getnet F, Bauer H, et al. Medium-term effects of conservation agriculture based cropping systems for sustainable soil and water management and crop productivity in the Ethiopian highlands. Field Crops Research. 2012;**132**:53-62

[67] Zhao X, Wu P, Persaud N. Soil quality indicators in relation to land use and topography in a small catchment on the Loess Plateau of China. Land Degradation & Development. 2013;**26**(1):54-61. DOI: 10.1002/ldr.2199

[68] Sparling GP. Ratio of microbial biomass carbon to soil organic carbon as a sensitive indicator of changes in soil organic matter. Australian Journal of Soil Research. 1992;**30**(2): 195-207

[69] Jacobs A, Rauber R, Ludwig B. Impact of reduced tillage on carbon and nitrogen storage of two Haplic Luvisols after 40 years. Soil and Tillage Research. 2009;**102**:158-164

[70] During RA, Thorsten H, Stefan G. Depth distribution and bioavailability of pollutants in long-term differently tilled soils. Soil and Tillage Research. 2002;**66**:183-195

[71] Meena JR, Behera UK, Chakraborty D, Sharma AR. Tillage and residue management effect on soil properties, crop performance and energy relations in greengram (*Vigna vadiate* L.) under maize-based cropping systems. International Soil and Water Conservation Research. 2015;**3**:261-272

[72] Mühlbachová G, Kusá H, Růžek P. Soil characteristics and crop yields under different tillage techniques. Plant Soil and Environment. 2015;**61**:566-572

[73] He Y, Xu M, Qi Y, Dong Y, He X, Li J, et al. Differential response of soil microbial community to four-decade long grazing and cultivation in a semi-arid grassland. Sustainability. 2017;**9**(1):128-142

[74] Govaerts B, Mezzalama M, Unno Y, Sayre KD, Luna-Guido M, Vanherck K, et al. Influence of tillage, residue management, and crop rotation on soil microbial biomass and catabolic diversity. Applied Soil Ecology. 2007;**37**:18-30

[75] Gajda AM, Przewłoka B. Soil biological activity as affected by tillage intensity. International Agrophysics. 2012;**26**:15-23

[76] Insam H, Goberna M. Use of Biolog® for community level physiological profiling (CLPP) of environmental samples. In: Kowalchuk GA, de Bruijn F, Head IM, Van der Zijpp AJ, van Elsas JD, editors. Molecular Microbial Ecology Manual. 2nd ed. Dordrecht: Kluwer Academic; 2004. pp. 853-860

[77] Habig J, Swanepoel C. Effects of conservation agriculture and fertilization on soil microbial diversity and activity. Environments. 2015;**2**:358-384

[78] Yang Q, Wang X, Shen Y. Comparison of soil microbial community catabolic diversity between rhizosphere and bulk soil induced by tillage or residue retention. Journal of Soil Science and Plant Nutrition. 2013;**13**:187-199

[79] Zhao D, Li F, Yang Q, Wang R, Song Y, Tao Y. The influence of different types of urban land use on soil microbial biomass and functional diversity in Beijing, China. Soil Use and Management. 2013;**29**:230-239

[80] Soman C, Li S, Wander MM, Kent AD. Long-term fertilizer and crop-rotation treatments differentially affect soil bacterial community structure. Plant and Soil. 2017;**413**:145-159. DOI: 10.1007/s11104-016-3083-y

[81] Zhang CC, Ke SS, Wang J, Ge Y, Chang SX, Zhu SX, Chang J. Responses of microbial activity and community metabolic profiles to plant functional group diversity in a full-scale constructed wetland. Geoderma. 2011;**160**(3-4):503-508. DOI: 10.1016/j.geoderma.2010.10.020

[82] Wei M, Tan F, Zhu H, Cheng K, Wu X, Wang J, et al. Impact of Bt-transgenic rice (SHK601) on soil ecosystems in the rhizosphere during crop development. Plant Soil Environment. 2012;**58**(5):217-223

[83] Birkhofer K, Schoning I, Alt F, Herold N, Klamer B, et al. General relationships between abiotic soil properties and soil biota across spatial scales and different land-use types. PLoS One. 2012;**7**(8):e43292

[84] Bevivino A, Paganin P, Bacci G, Florio A, Pellicer MS. Soil bacterial community response to differences in agricultural management along with seasonal changes in a mediterranean region. PLoS One. 2014;**9**(8):e105515

[85] Shange RS, Ankimah RO, Ibekwe AM, Zabawa R, Dowd SE. Distinct soil bacterial communities revealed under a diversely managed agroecosystem. PLoS One. 2012;**7**:e40338

[86] Bhuyan SI, Triathi OP, Khan ML. Effect of season, soil and land use pattern on soil N-mineralization, ammonification and nitrification: A study in Arunachal Pradesh, eastern Himalaya. International Journal of Environmental Science. 2014;**5**(1):88-97

[87] Liu L, Qin S, Lu D, Wang B, Yang Z. Variation of potential nitrification and ammonia oxidizing bacteria community with plant growing period in apple orchard soil. Journal of Integrative Agriculture. 2014;**13**(2):415-425

[88] Birkhofer K, Bezemer TM, Bloem J, Bonkowski M, Christensen S, et al. Long-term organic farming fosters below and aboveground biota: Implications for soil quality, biological control and productivity. Soil Biology and Biochemistry. 2008;**40**:2297-2308

[89] Sipilä TP, Yrjälä K, Alakukku L, Palojärvi A. Cross-site soil microbial communities under tillage regimes: Fungistasis and microbial biomarkers. Applied and Environmental Microbiology. 2012;**78**(23):8191-8201

[90] Bissett A, Richardson AE, Baker G, Thrall PH. Long-term land use effects on soil microbial community structure and function. Applied Soil Ecology. 2011;**51**:66-78

[91] Natywa M, Sawicka A, Wolna-Maruwka A. Microbial and enzymatic activity in the soil under maize crop in relation to differentiated nitrogen fertilisation. Water-Environment-Rural Areas. 2010;**2**(30):111-120

[92] Mahajan GR, Manjunath BL, Singh NP, Ramesh R, Verma RR, Latare AM, et al. Effect of organic and inorganic sources of nutrients on soil microbial activity and soil organic carbon build-up under rice in west coast of India. Archives of Agronomy and Soil Science. 2016;**63**(3):414-426. DOI: 10.1080/03650340.2016.1213813

Chapter 4

Sustainable Agroecosystems for Livelihood Security in Indian Himalayas

Ram Prakash Yadav, Jaideep Kumar Bisht,
Vijay Singh Meena and Mahipal Choudhary

Additional information is available at the end of the chapter

http://dx.doi.org/10.5772/intechopen.74495

Abstract

Agroforestry systems are an age-old practice in the Indian Himalayan region. Agroforestry deals with the combination of tree species with crop plants, fisheries, animals, bee keeping, and so on, and it is based on the principle of optimum utilization of land. Agrihorticulture, silvihorticulture, hortipastoral, and silvipastoral systems are diversified land use options for agroforestry in the hill region. The study was conducted at experimental farm Hawalbagh (29°36′N and 79° 40′ E, 1250 m amsl) of Vivekananda Parvatiya Krishi Anusandhan Sansthan, Almora, India. Study in an agri-horti system revealed that ragi (*Eluesine coracana*) and soybean (*Glysine max*) during *kharif* (rainy season) and wheat (*Triticum estivum*) and lentil (*Lens esculenta*) during *rabi* (winter season) can be grown successfully with pecan nut (*Carya illinoinensis*) tree without significant reduction in the yield of the crop. However, grain yield of these crops was numerically higher in the field without pecan nut tree. In fruit-based agri-horti system four fruit crops, hill lemon (*Citrus limon*), pear (*Pyrus communis*), plum (*Prunus domestica*), and apricot (*Prunus armeniaca*) were planted with soybean in *kharif* and dual purpose wheat during *rabi* season. During initial years, no significant effect on grain yield was observed with the presence of different fruit trees. Green forage yield varied from 4600 to 5900 kg/ha in different treatments. In different treatments, ginger (*Zingiber officinale*) and turmeric (*Curcuma longa*), turmeric and taro (*Colocasia esculenta*), and two varieties of turmeric (Pant Pitabh and Swarna) were grown under *Grewia optiva, Quercus leucotrichophora, Bauhinia variegata*, and *Celtis australis*. Turmeric and ginger produced significantly higher yield (12.04 and 7.99 t/ha) under oak. The highest rhizome yield was obtained under *Quercus leucotrichophora* (11,738 kg/ha) followed by *Bauhinia variegata*. Pant Pitabh gave significantly higher yield (10,860 kg/ha) than swarna. Improved systems with five tree species, that is, *Grewia optiva, Quercus leucotrichophora, Bauhinia retusa, Melia azedarach*, and *Morus alba* and four grasses, that is, *Setaria kazugulla, Setaria nandi*, Congo signal, and Broad leaf paspalum (*Paspalum* spp.) were tested under the silvipastoral system. *Quercus leucotrichophora* yielded (10,675 kg/ha) significantly higher green biomass than others, and the lowest green biomass was harvested from *Grewia optiva*. Among grasses, *Setaria nandi* produced the highest green forage (6234 kg/ha). Thus, in hilly terrain, planting of inter-spatial woody perennials, with least negative influences on the agronomic crops, seems productive in agroforestry system for settled farming. Therefore, agroforestry is a set of land use alternative, which if developed for resource poor farmers, can provide increased values and reduced risks and it should be made more popular in the rural areas.

Keywords: agroforestry systems, Himalaya, crop yield, agrihorticulture, woody perennials, silvihorti, silvipastoral, sustainability

---

1. Introduction

Agroforestry systems are an age-old practice in the Indian Himalayan region. Agroforestry deals with the combination of tree species with crop plants, fisheries, animals, bee keeping, and so on, and it is based on the principle of optimum utilization of land. Agrihorticulture, silvihorticulture, hortipastoral, and silvipastoral systems are diversified land use options for agroforestry in the hill region. These agroforestry systems play a vital role in the livelihood of the hill people [1, 2]. The production and availability of green fodder is not uniform throughout the year, available in plenty during monsoon and remaining period, that is, winter and summer are miserable in hilly areas of Himalaya. The leaves of some identified evergreen fodder trees are given to animals during acute winter period [3]. *Grewia optiva, Quercus leucotrichophora, Quercus dilatata, Quercus semecarpifolia, Ficus nemoralis, Ficus palmata, Ficus roxburghii,* and so on are the main winter fodder trees [1, 2].

In the Himalayan region, several indigenous agroforestry systems based on people's needs and site-specific characteristics have been developed over the years. Agroforestry practices have wide and promising potential to store carbon and remove atmospheric carbon dioxide through enhanced growth of trees. However, little has been reported regarding carbon sequestration potential of agroforestry systems in Indian Himalaya. In this scenario, it is imperative that the carbon sequestration potential for agroforestry practices in the region is also investigated. The structural and functional aspects of tree species in traditional agroforestry systems greatly affect the overall productivity of the system. Generally, the overall productivity (crops + trees) in agroforestry systems is higher than that in sole cropping systems [4]. Nair [5] defined agro-forestry as "a land use system that involves deliberate retention, introduction or mixture of trees or other woody perennials in crops and animal production to benefit from the resultant ecological and economical intersections. Hence, the basic objective of the study is (1) what are the land uses for improved farm production, income generation, and livelihood security (2) to investigate ecosystem services of these land uses, which are maintaining their sustainability.

1.1. Area and climate

The Indian Himalayas cover an area of 53.7 Mha, which ~17% of total geographical area of the country. Out of 21 agro-ecological regions of the country, four regions are enclosed entirely and one partially in the hilly agro-ecosystem. All five agro-ecological zones surround a broad distinction in their environment ranging from cold arid to warm per humid. The mean annual precipitation in the area ranged between 150 and 4000 mm. The average annual temperature fluctuated between 8 and 22°C. The yearly variation in cultivation phase of diverse crops varies from 90 to ~270 days. The major soils groups of the region are skeletal and calcareous to brown forest podzolic [6], which are alkaline and acidic by nature. The natural flora is alpine, temperate, wet evergreen and tropical deciduous forest in the Indian Himalayan region.

1.2. Land utilization and population

In the Himalayan region, population is ~39 million, which is ~4% of the total population of the India; the primary source of income here is agriculture, which contributes ~45% of the total regional income.

Himalayan area is sparsely inhabited (627 people/1000 ha). However, in this region, the genuine stress on farming land is much high as ~15% of the reporting area is net cultivated area. Availability of land for cultivation in hilly region of Himalaya is just ~ 0.17 ha per capita. In the Himalayas, woodland covers ~ 47% of total reporting area of hills, and it is the main land use, whereas ~ 13% area of the region comes under permanent pastures and grazing lands.

Horticulture has its advantage in the Himalayan ecosystem due to its particular ecological condition and numerous micro-situations. This region has a subtropical to temperate climatic condition, and the broad choice of fruits, such as apple (*Malus domestica*), banana (*Musa bauensis*), pomegranate (*Punica granatum*), citrus (*Ctrus* spp.), mango (*Mangifera indica*), peach (*Prunus persica*), plum (*Prunus domestica*), and walnut (*Juglans regia*); vegetables, such as brinjal (*Solanum melongena*), cabbage (*Brassica oleracea var. capitata*), cauliflower (*Brassica oleracea var. botrytis*), garlic (*Allium sativum*), onion (*Allium cepa*), pea (*Pisum sativum*), potato (*Solanum tuberosum*), tomato (*Solanum lycopersicum*), and so on; and spices, such as chilies (*Capsicum annum*), ginger (*Zingiber officinale*), and turmeric (*Curcuma longa*) is cultivated [2]. According to Tulachan [6], under horticulture (fruits and vegetable), ~ 13–15% is the gross cropped area.

1.3. The forests

The non-arable land occupies ~84% out of which forest land constitutes ~ 60% of the total geographical area [7]. Forest plays a vital role to the sustenance of agrarian economy, and inhabitants withdraw resources such as forage material, fuelwood, non-timber forest product, and livestock grazing. These forests on account of heavy pressure are under degradation at a rapid pace. In the Himalayan region, agriculture has been convoluted integration with forestry practices. However, the increased availability of fodder, fuelwood, and minor forest products including ecological security, larger driving force is needed to put on agroforestry [1].

2. Agroforestry systems

Agroforestry is intentional incorporation of agriculture, forestry, animals and/or pastures, which are managed intensively and in an integrated manner and each component interacts with each other to obtain ecological, monetary, and societal benefits. It is a science that provides information on the impact of the variables such as tree species, tree-spacing, orientation of trees, rotation of crops, pruning types and intensity, quantity and timing of litter addition, thinning and felling, which can be managed by the farmer. Trees not only supplement the fodder, fuel, fiber, fruits, and so on, but also accumulate biomass and sequester carbon [1, 2, 8, 9]. Dhyani *et al.* [10] asserted that agroforestry is key path to prosperity for millions of farm families leading to extra income, employment generation, greater food and nutrient security, and meeting other basic human needs in a sustainable manner. It is more pertinent to hilly regions where existence without agroforestry is difficult.

2.1. Objectives of agroforestry

- To utilize available farm resources properly
- To maximize per unit production of food, fodder and fuel
- To optimize biological and physiological resources
- To maintain ecological balance
- To check soil erosion, conserve soil moisture, and increase soil fertility

2.2. Identification of agroforestry systems

Structural (nature, arrangement) and functional (outputs role) basis [11] was adopted for identification and naming agroforestry systems prevalent in the study area. System types

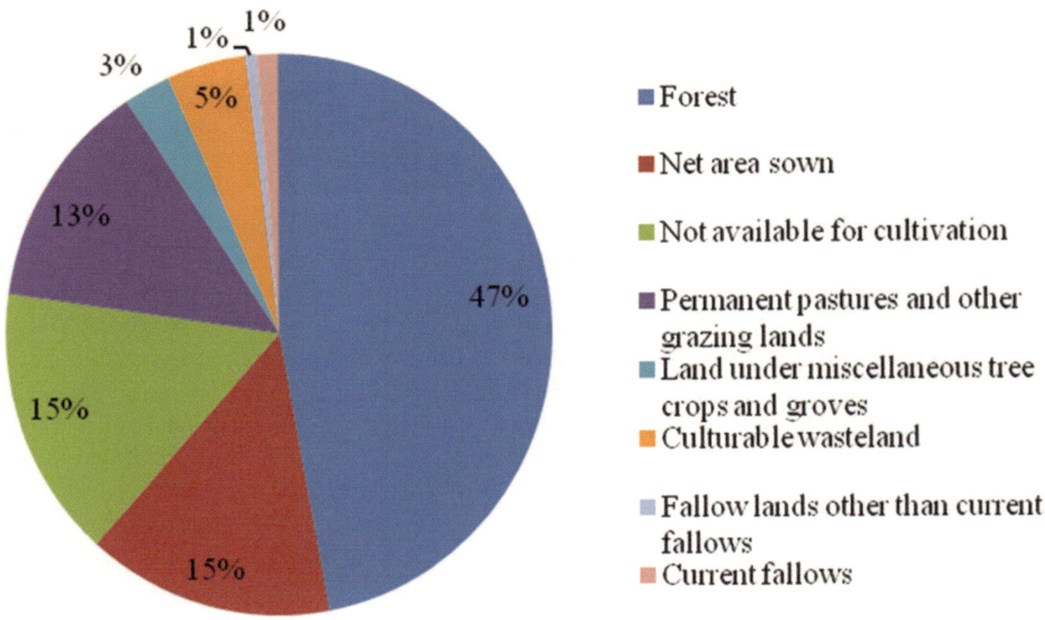

Figure 1. Land utilization pattern of North-West (Uttarakhand, Himachal Pradesh and Jammu & Kashmir) Himalaya.

and their system units were indicated using stratified classification of agroforestry practices as given by Zou and Sanford [12]. Considering the major components, a system type was named, whereas functional unit, that is, combination of specific tree species or other related components, was termed system unit. Hence, functional units such as cereals, pulses and vegetable in agriculture; in horticulture, specific fruit trees; in grasslands, grasses; and in silvipasture, trees/grass species were considered to identify systems and system units.

The different components of agroforestry systems were identified as primary and secondary. (1) Primary components: The components occupying the larger area of the total unit area and serving the major function, that is, production of primary output needed by the farmers was termed as primary component. (2) Secondary components: The component occupying relatively lesser area of the total unit area compared to area under primary component and yielding secondary needs of the farmers was termed as secondary component (**Figure 1**). Using abovementioned framework for identifying agroforestry systems, it can be classified as follows:

3. Agroforestry ecosystem services

3.1. Soil conservation

From the land a lot of losses of valuable is taking place in nature. These losses are in terms of water loss and soil loss through runoff and sedimentation, and precious top productive soil is being lost in a huge amount in this way. Estimates indicated that soil is displaced annually ~16 tones/ha which washed into the sea, which is greater than the permitted limit of 4.5 tones/ha/year. Agroforestry can be used to reduce this loss through putting it on barren lands of a watershed in the form of the land cover. Total annual accretion of litter fall in pecan nut (6 m × 7 m) based agri-horti system found 2.14 t ha^{-1} $yr.^{-1}$ and relative abundance of nutrients in litter fall of pecan nut tree were in the order of C (901.9 kg ha^{-1} $yr.^{-1}$) > N (57.44 kg ha^{-1} $yr.^{-1}$) > K (43.29 kg ha^{-1} $yr.^{-1}$) > P (3.21 kg ha^{-1} $yr.^{-1}$), which are supposed to help in nutrient buildups of the soil (**Figure 2** and **Table 1**) [13, 14].

3.2. Microclimate

Agroforestry will help in the moderation of microclimate around the trees. Microclimate amelioration, which involves air and soil moisture and temperature relations, results primarily from the use of trees in shade and live fences, growing crops in interspaces of the trees, creating shelterbelt and/or windbreak, and boundary plantation of trees. The condition of shade creates a net effect through complex interactions, and it expands further than the simply light and heat moderation [14]. Humidity, temperature and air movement as well as soil temperature and moisture of the soil directly influence transpiration, photosynthesis, and balance of energy of associated crops [15], and the overall influence may be translated into improved yields. The numerous practices that farmers conventionally have devised to achieve this objective confirm to the significance ascribed to microclimate moderation [16–18].

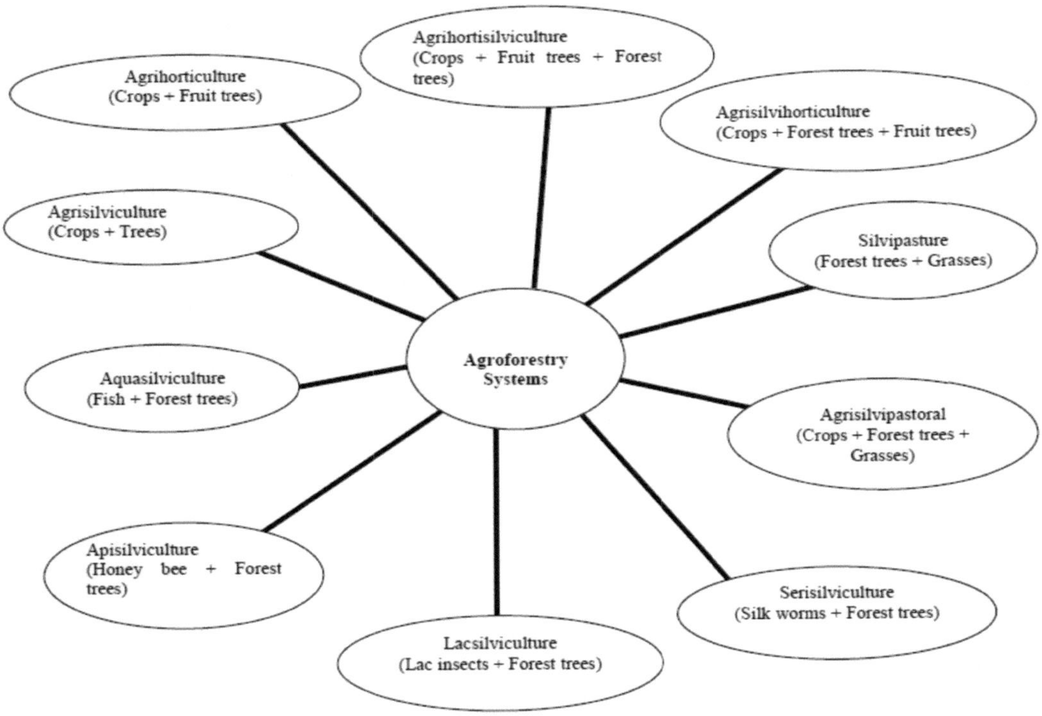

Figure 2. Classification of agroforestry system based on structure and function [8].

Nutrients	Concentration (%)	Potential nutrient return (kg ha^{-1} yr^{-1})
Carbon	42.1	901.9
Nitrogen	2.68	57.4
Phosphorus	0.15	3.2
Potassium	2.02	43.2

Table 1. The average concentration of nutrients in litter fall of pecan nut tree and potential amount of nutrients that could be released from pecan nut litter fall.

3.3. Watershed management

The watershed area can be utilized properly with adoption of agroforestry. This approach of watershed decides how to manage the land, that is, plan of crop management, development of pasture, plantations of forest, and so on, as per need of the watershed. Watershed management includes judicious use of natural resources, that is, water, land, plants and animals including human for conservation and regeneration purpose within the watershed area. As it is the human who is principally accountable for environmental degradation, this management of watershed maintains balance in the environment between natural resources including man and animals that inhabit the watersheds. Measures such as employment of people, planning from bottom-up approach, sustainability and equitable distribution of resources are the approaches for a good watershed management.

Approaches such as in situ conservation measures for soil and moisture, that is, bunding, vegetative barriers, terrace, trenching, treatment of drainage line by vegetative and engineering structures and developing small structures for water harvesting. Plantation of multi-purpose trees (MPTs), legumes, shrubs and sowing of grasses for grazing land development. In the watershed area promoting natural regeneration, encouraging creation of agroforestry & horticulture and establishment of woodlots to conserve common property resources and to meet demand of fuelwood are the activities need to promote.

3.4. Biomass production and carbon storage

The maximum production of biomass per unit area is a main objective of agroforestry. For the production of biomass, we can use the bunds, canal roads, farm ponds, lakes, waterlogged area and ravines of a watershed in an economic way. Aboveground biomass production and carbon storage (**Figure 3**) in some of the important agroforestry systems viz., fruit tree-based agrihorticulture, oak high-density plantation (1 × 1 m), and pecan nut based agrihorticulture system was studied by Yadav et al. [9, 19–21]. Due to population explosion and increasing number of industries, pressure on forest is increasing day by day. Because of this population pressure, it is not possible to grow forest on agricultural lands. Through agroforestry, we can grow trees on marginal lands with crops. This will also restore the ecological balances.

3.5. Livelihood security

In watershed area, agro-based cottage industry can be promoted with the help of agroforestry. These will include paper pulp industry, herbal medicines, fiber production, piggery, poultry, aquaculture, beekeeping, dairying, sericulture, lac-culture and mushroom production. It will help in generating income and improving living standard of marginal farmers.

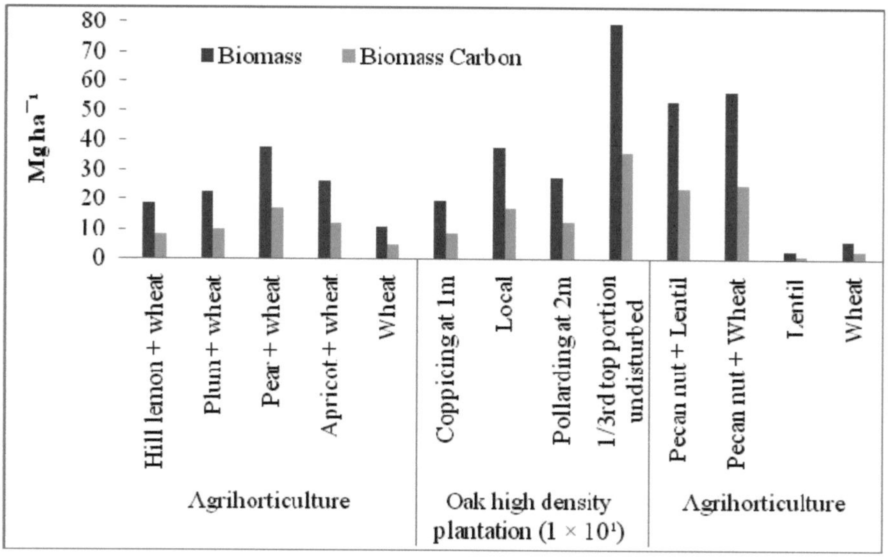

Figure 3. Aboveground biomass and biomass carbon in different systems of Central Himalaya [9, 19–21].

In Indian Himalaya, livelihood security is reliant on traditional farming practices that include agroforestry beside forest produce; forests provide 73–79% of required energy from fuelwood and more than 81% fodder [2]. The income of people was positively correlated with livestock rearing and farming. The adoption of this type of integrated farming which includes agroforestry with livestock (dairy and poultry/goat rearing) will improve the income of the people and sustainability of farming. Awareness of agroforestry was significantly related to literacy rate, land holding size, and fuelwood consumption (**Table 2**). A significantly positive relationship was obtained among income, fodder consumption, fodder from forest, land holding size, and livestock size [2].

3.6. Climate change mitigation

The mitigation actions deal with the causes of the crisis ever-increasing greenhouse gases concentrations. Example includes reduction of energy consumption and promoting clean technologies could be mitigation measures. Mitigation is an international issue, as when a mitigation project related to agroforestry reduces emissions of greenhouse gases, it remunerates the entire globe. Many agroforestry activities contribute to climate change mitigation (**Figure 4**). C stocks can be increased by creating agroforestry through plantations [3]. As shown in the graph, the difference between the present growing stock and the baseline through creating plantations of agroforestry is a benefit. Reduced felling or no felling at all of trees is helpful to conserve existing growing stocks on agroforestry farm lands. With reference to the degradation or tree felling scenario, the benefit of conservation is estimated in this situation. In agroforestry activities, emissions can be reduced, with a reduction of energy or fertilizers in diverse agroforestry operations. From agroforestry products, biomaterials and

Parameter	Correlation									
	1	2	3	4	5	6	7	8	9	10
Annual income	0.703**	1								
Livestock size	0.865**	0.488**	1							
Fuel consumption	0.174	0.273	0.179	1						
Fodder from forest	0.551**	0.313	0.785**	0.307	1					
Family size	−0.132	0.007	−0.006	−0.020	0.033	1				
Sex ratio	0.181	0.032	0.151	0.240	0.192	−0.201	1			
Literacy rate	0.176	0.340	0.056	0.361	0.024	0.016	0.354	1		
Land holding size	0.530**	0.378*	0.447*	0.235	0.403*	−0.467**	0.495**	0.432*	1	
Agroforestry awareness	0.168	0.325	0.058	0.375*	0.026	0.000	0.352	0.998**	0.434*	1

Note: **Correlation significant at the 0.01 level. *Correlation significant at the 0.05 level.

Table 2. Pearson's correlation coefficients between various parameters of Central Himalayan watershed [2].

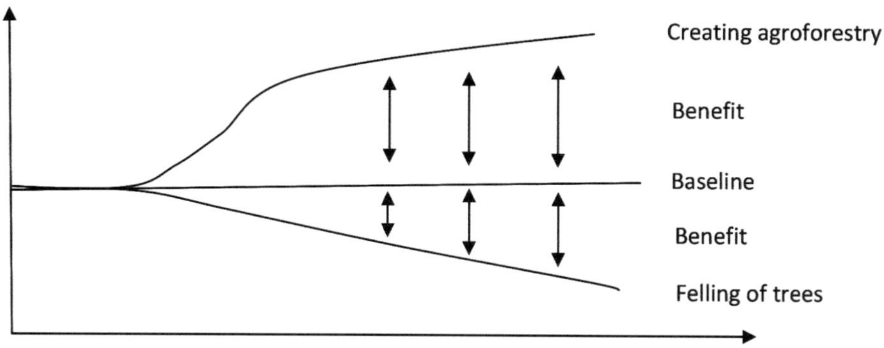

Figure 4. Carbon sequestration through creating agroforestry plantation and reduced felling of trees from agroforestry farmlands [8].

bioenergy can be produced to substitute materials or energy that generates GHGs. The first two activities refer to C sequestration in the agroforestry or any other ecosystem, while the last two refer to energy-related emissions [8].

4. Prevalent agroforestry system in the Himalayan region

Several agroforestry systems are common in the hill region [22]. Growing trees on the field bunds and around the fields is an age-old practice. Agrihorticulture, silvihorticulture, horti-pastoral and silvipastoral systems are diversified land use options for agroforestry in the hill region [2]. These agroforestry systems play a vital role in the livelihood of the hill people [1, 2]. The study on some important agroforestry systems was conducted at experimental farm Hawalbagh (29°36′N and 79°40′E, 1250 m amsl) of Vivekananda Parvatiya Krishi Anusandhan Sansthan, Almora, India. The most common agroforestry systems in hills are as follows:

4.1. Agrihorticulture system

Agrihorticulture systems are the most common systems, which are pervasive in Himalaya [2]. These systems (agriculture crops + fruit trees) are the combination of agriculture crops grown in the interspaces of fruit trees. These systems are the backbone of food, nutritional and livelihood security in hilly terrain. Though people do not follow specific spacing and orientation patterns for planting fruit trees but mainly aimed to obtain diversified outputs, these systems improved farm income and reduced risk.

Study in an agrihorticulture system revealed that ragi and soybean during *kharif* and wheat and lentil during *rabi* can be grown successfully with pecan nut tree without significant reduction in the yield of the crop [22]. However, grain yield of these crops was numerically higher in the field without pecan nut tree. In fruit-based agrihorticulture system, four fruit crops, hill lemon, pear, plum and apricot were planted with soybean in *kharif* and dual purpose wheat during *rabi* season. During initial years, no significant effect on grain yield was observed with the presence of different fruit trees (**Table 3**).

Crop sequence	Grain yield (t ha^{-1})	
With pecan nut tree	*Rabi*	*Kharif*
Maize-pea	1.56	3.19
Soybean-wheat	2.96	1.50
Maize-wheat	3.00	3.47
Soybean-pea	1.82	1.62
Without pecan nut tree		
Maize-pea	1.75	3.75
Soybean-wheat	3.32	1.56
Maize-wheat	3.10	3.49
Soybean-pea	1.59	1.48

Table 3. Yield of field crops with and without pecan nut tree (average of 5 years).

4.2. Silvihorticulture system

Silvihorticulture systems (fodder trees + horticulture crops) are the combinations of the fodder trees and horticulture plantation crops. Fodder trees serve the purpose of forage supply to livestock, which are paramount to rural agrarian economy. Beside forage supply, these trees also fulfill energy (fuelwood) needs in the form of the rural inhabitants of the hilly region. Livestock rearing and horticulture plantations crops generate income to farming community.

Green forage yield varied from 4600 to 5900 kg/ha in different treatments in study on silvihorticulture. In different treatments, ginger and turmeric, turmeric and taro, and two varieties of turmeric (Pant Pitabh and Swarna) were grown under *Grewia optiva, Quercus leucotrichophora, Bauhinia variegata,* and *Celtis australis* [23]. Turmeric and ginger produced significantly higher yield (12.04 and 7.99 t/ha) under oak. The highest rhizome yield was obtained under *Quercus leucotrichophora* (11,738 kg/ha) followed by *Bauhinia variegata*. Pant Pitabh gave significantly higher yield (10,860 kg/ha) than swarna (**Table 4**).

4.3. Silvipastoral system

Silvipastoral systems (fodder trees + grasses) are managed on the barren and marginal land including community wasteland. In this system, both components provide fodder materials to livestock owners. Dairying industry helps farmers to fetch good income through selling milk and its byproducts. Draught animals are source of energy for agricultural and other domestic activities.

Improved systems with five tree species, that is, *Grewia optiva, Quercus leucotrichophora, Bauhinia retusa, Melia azedarach,* and *Morus alba* and four grasses, that is, *Setaria kazugulla, Setaria nandi*, Congo signal, and Broad leaf paspalum (*Paspalum* spp.) were tested under silvipastoral system [3]. *Quercus leucotrichophora* yielded (10,675 kg/ha) significantly higher green biomass than others, and the lowest green biomass was harvested from *Grewia optiva*. Among grasses, *Setaria nandi* produced the highest green forage (6234 kg/ha) (**Table 5**).

4.4. Hortipastoral system

For proper management of marginal land, the fruit trees are planted depending upon agro-climatic situations. In unutilized interspaces of the fruit trees grasses are cultivated to produce green forage to feed animals. In hortipastoral systems (fruit trees + grasses), one component sold directly in the market to generate income, whereas other component fulfills forage needs of the livestock. This system helps in generating income and improving living standard of rural populations.

In hortipastoral production system, four *rabi*, namely, perennial rye (*Lolium perenne*), tall fescue (*Festuca arundinacea*), hima 14 (*Festuca arundinacea var. hima)*, and grassland manava (*Festuca jatia*) grasses with and without peach tree were grown for utilization of marginal land. Grassland manava produced significantly higher mean green forage yield (284.1 q ha^{-1}) fol-

Treatment	Yield (t ha^{-1})		
	1994–1995	1995–1996	Average
Turmeric			
Quairal	6.03	4.66	5.35
Kharik	12.77	8.57	10.67
Oak	15.00	9.08	12.04
Bhimal	7.70	5.70	6.70
Open	10.75	7.16	8.96
Ginger			
Quairal	5.78	3.15	4.47
Kharik	10.44	4.95	7.70
Oak	10.74	5.23	7.99
Bhimal	6.67	3.38	5.03
Open	9.12	3.91	6.52

Table 4. Rhizomes yield of ginger and turmeric with different fodder trees.

Grass species	Green forage yield (t ha^{-1})	
	Pine	Deodar
Pangola	14.01	13.66
Rhodes	4.78	3.06
Para grass	0.48	2.04
Guine	0.41	2.55
Local	0.25	0.22

Table 5. Performance of different improved and local grasses under pine and deodar.

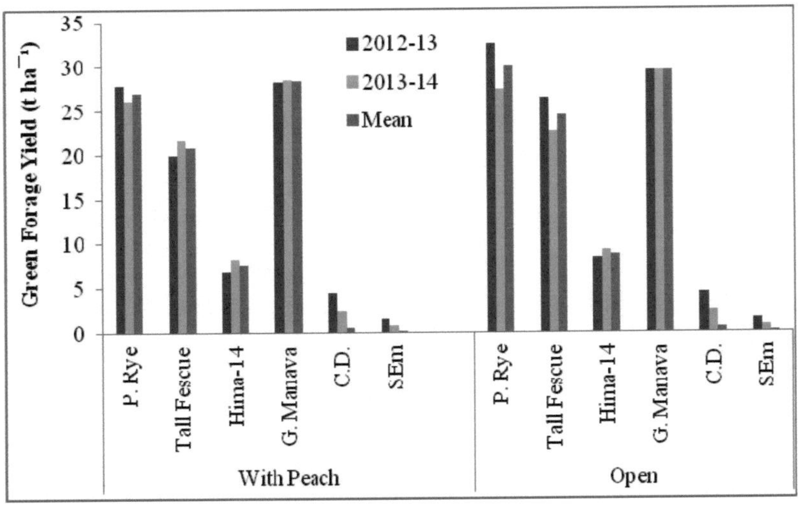

Figure 5. Green forage yield (q ha^{-1}) of winter grasses in peach-based hortipastoral production system [24].

lowed by perennial rye under peach tree, while in open perennial rye gave higher mean green forage yield 299.6 q ha^{-1}, which was at par with the grassland manava 294.7 q ha^{-1}(**Figure 5**) .

4.5. Energy plantation system

This system not only restores and conserves community lands but also improves its productivity through production of timber, fodder, and fuelwood. Energy plantation has been found a viable agroforestry system for the hills. For such system, Quairal (*Bauhinia variegeta*), Bhimal (*Grewia optiva*), Batain (*Melia azedarach*), oak (*Quercus leucotrichophora*), kharik (*Celtis australis*), *Bedu (Ficus palmata*), and Paiya (*Prunus padam*) have been found suitable for energy plantation [25].

The green matter yields [25], however varied with the varying tree species recorded on an average, 50–250 q/ha in 3rd year and 400–650 q/ha in the subsequent years. In case of oak energy plantation, combined application of chemical fertilizers (30 kg N + 40 kg P_20_5/ ha) with manure (FYM @ 10 t/ha) produced 14.0% more leaf yield than chemical fertilizer (30 kgN + 40 kg P_20_5) alone (12.4 tons/ha) and 11% higher than manure alone (12.7 tons/ha).

4.6. Agroforestry in cold arid zone

In Himalayan cold arid zone, agroforestry systems need to be stressed as follows:

i. Agrisilviculture: Here legume plants are cultivated with densely planted poplar (*Populus nigra*), and it is well accepted in Ladakh region of Indian Himalaya [26].

ii. Silvipastoral: such agroforestry systems are dominated as pastures at high altitude of Ladakh Caragana, Hippophae (*Hippophe rhamnoides*) and willows (*Salix alba*) combinations are practiced. The junipers, birch and rhododendron are found in alpine zone of Spiti with Poa and Agropyron spp. of grasses are cultivated [27].

iii. Agrisilvipastoral: Crops like wheat or barley is grown in the interspaces of *Populus*, *Salix* and *Robina* spp. and lopped material are used as animal fodder.

iv. Hortipastoral: In alfalfa field, fruit plants such as apricot or apple are grown.

Thus, in hilly terrain, planting of interspatial woody perennials, with least negative influences on the agronomic crops, seems productive through agroforestry system for settled farming. Instead of this, these systems provide environmental benefits such as soil improvement and carbon sequestration [19, 28]. Therefore, agro forestry is a set of land use alternative, which, if developed for resource poor farmers, can provide increased values and reduced risks, and it should be made more popular in the rural areas. The mixed farming and alley cropping system appears to be viable alternatives for farmers practicing settled cultivation in hills' base and plain areas, while in hilly terrain, planting of interspatial woody perennials on terrace risers, with least negative influences on the agronomic crops, seems productive through agroforestry system for settled farming.

5. Conclusion

Agroforestry is a set of land use alternative, which, if developed for resource poor farmers, can provide increased values and reduced risks. This will help the poor farmer in overcoming the fuel and fodder shortage with increased productivity of the land and pressure on forest will be reduced. This will generate income to farmer in a drought year when crop is failed. This will have a positive effect on the environment as it reduces erosion and runoff, improves soil fertility, and mitigates climate change. Therefore, agro forestry should be made more popular in the rural areas through extension programs, and a top priority should be given in the watershed development.

Acknowledgements

The authors are grateful to the Director, Indian Council of Agricultural Research- Vivekananda Parvatiya Krishi Anusandhan Sansthan, Almora, Uttarakhand for providing necessary support.

Conflict of interest

The authors hereby declare that there is no conflict of interest.

Author details

Ram Prakash Yadav*, Jaideep Kumar Bisht, Vijay Singh Meena and Mahipal Choudhary

*Address all correspondence to: rams34052@gmail.com

ICAR-Vivekananda Parvatiya Krishi Anusandhan Sansthan, Almora, India

References

[1] Yadav RP, Bisht JK. Agroforestry: A way to conserve MPTs in North Western Himalaya. Research Journal of Agriculture and Forestry Sciences. 2013;**1**:8-13

[2] Yadav RP, Gupta B, Bhutia PL, Bisht JK. Socioeconomics and sources of livelihood security in Central Himalaya, India a case study. International Journal of Sustainable Development & World Ecology. 2016;**23**:1-9. DOI: 10.1080/13504509.2016.1239233

[3] Bisht JK, Yadav RP. Sustainable fodder production strategy through utilization of wastelands in hills. Book of Abstracts, 3rd World Congress on Agroforestry, Trees for life–Accelerating the impact of agroforestry. Nairobi: World Agroforestry Centre. 2014;24. www.worldagroforestry.org/downloads/Publications/PDFS/B17335.PDF

[4] Chaturvedi AN. Poplar for planting. In: Uttar Pradesh Department Bull. India. 1981;no-50 Lucknow 27

[5] Nair PKR. Soil Productivity Aspect of Agroforestry. Nairobi, Kenya: ICRAF; 1984

[6] Tulachan PM. The state of mountain agriculture in India. In: Rana GM, Shrestha B, Joshi SM, editors. State of Mountain Agriculture in the Hindu Kush-Himalayas: A Regional Comparative Analysis. Kathmandu: ICIMOD; 2001. pp. 25-31

[7] Alam A. Long-term Strategies and Programmes for Mechanization of Agriculture in Agro-Climatic Zone–I: Western Himalayan Region. 2016;78-86. http://farmech.gov.in/06035-04-ACZ1-15052006.pdf

[8] Yadav RP, Bisht JK, Gupta B, Mondal T. Toward the sequestration potential of agroforestry practices to combat climate change in Kumaon Himalaya, India. In: Bisht JK, Meena VS, Mishra PK, Pattanayak A, editors. Conservation Agriculture. 2016;293-313. DOI: 10.1007/978-981-10-2558-7_11

[9] Yadav RP, Bisht JK, Bhatt JC. Biomass, carbon stock under different production systems in the mid hills of Indian Himalaya. Tropical Ecology. 2017;**58**(1):15-21

[10] Dhyani SK, Kareemulla KA, Handa AK. Agroforestry potential and scope for development across agro-climatic zones in India. Indian Journal of Forestry. 2009;**32**(2):181-190

[11] Nair PKR. Agroforestry systems inventory. Agroforestry Systems. 1987;**5**:301-317

[12] Zou X, Sanford RLJ. Agroforestry system in China: Survey and classification. Agroforestry Systems. 1990;**11**:85-94

[13] Yadav RP, Bisht JK. Litter fall and potential nutrient returns from pecan nut (*Carya illinoinensis*) in agroforestry system in Indian Himalaya. International Journal of Herbal Medicine. 2014;**2**:51-52

[14] Willey RW. The use of shade in coffee, cocoa and tea. Horticultural Abstracts. 1975;**45**: 791-798

[15] Rosenberg NJ, Blad BL, Verma DB. Microclimate: The Biological Environment. 3rd ed. New York, USA: Wiley & Sons; 1983

[16] Wilken GC. Microclimate management by traditional farmers. Geographical Review. 1972;**62**:544-566

[17] Stigter CJ. Microclimate Management and Manipulation in Traditional Farming. CAgM Report No. 25. Geneva, Switzerland: World Meteorological Organization; 1988

[18] Reifsnyder WS, Darnhofer TO, editors. Meteorology and Agroforestry. Nairobi, Kenya: ICRAF;1989

[19] Yadav RP, Bisht JK, Pandey BM. Aboveground biomass and C stock of fruit tree based land use systems in Indian Himalaya. Ecoscan. 2015;**9**:779-783

[20] Yadav RP, Bisht JK, Pandey BM, Kumar A, Pattanayak A. Cutting management versus biomass and stock of oak under high density plantation in central Himalaya, India. Applied Ecology and Environmental Research. 2016;**14**:207-214

[21] Yadav RP, Bisht JK, Mondal T and Mishra PK. Agroforestry practices for sequestration in the mid hills of central Himalaya. In: Peshin R, Dhawan AK, Bano F, Resam KS, editors. The Proceeding of the Natural Resource Management: Ecological Perspectives, Indian Ecological Society International Conference-2016, SKUAST-Jammu. 2016;477

[22] Bisht JK, Srivastva AK. Alternate land uses and agro-forestry systems in hills. In: Gupta HS, Srivastva AK, Bhatt JC, editors. Sustainable Production from Agricultural Watersheds in NWH. VPKAS. 2006;441-456

[23] Bisht JK, Chandra S, Chauhan VS, Singh RD. Performance of ginger (*Zingiber officinale*) and turmeric (*Curcuma longa*) with fodder tree based silivi-horti system in hills. Indian Journal of Agricultural Science. 2000;**70**(6):578-582

[24] Yadav RP, Bisht JK, Mondal T, Pattanayak A. Forage Production in Peach Based Hortipastoral System in Indian Himalaya. International Grassland Congress-2015 held at New Delhi from 20th–24th November, 2015

[25] Bisht JK, Srivastva AK, Gupta HS. Sustainable fodder production management in NW Himalaya. Technical Bulletin. Almora: VPKAS. 2009;**32**(2):1-62

[26] Sharma JP, Mir AM. Dynaimcs of Cold Arid Agriculture. Ludhiana, India: Kalyani Publishers

[27] Gupta RD, Sharma PD and Verma KS. 2000. Environment and ecology of cold arid region – A case study of Lahaul and Spiti in Himachal Pradesh. In: Sharma, Mir, editor. Dynamics of cold arid agriculture. Ludhiana, India: Kalyani Publishers;2000

[28] Yadav RP, Bisht JK. Agroforestry to combat climate change. Kheti. 2014;**66**:11-14. (In Hindi)

Chapter 5

The Cacao Agrosystems in Tabasco, México

Hortensia Brito-Vega,
Edmundo Gómez-Méndez and
José Manuel Salaya-Domínguez

Additional information is available at the end of the chapter

http://dx.doi.org/10.5772/intechopen.78302

Abstract

The cacao agrosystems are very suitable for reforesting completely cleared areas and can become biological corridors between segments of the forest, allowing the repopulation of birds, mammals, reptiles, and amphibians, among others. Cacao (*Theobroma cacao* L.) is one of the most important tropical crops both nationally and internationally. However, by appropriate management of cacao crops and the inclusion of aggregate values in forest, fruit, vegetables, and ornamental production, as organic cultivation, specific varieties of plants could generate significant income for small producers in the Southeast from Mexico. This cacao agrosystem is involved in erosion, soil fertility, plant nutrition, water quantity and quality, carbon sequestration, reduction of greenhouse gas emissions, and soil biodiversity.

Keywords: agroforestry, nutrient, organic, soil chemical properties

1. Introduction

Currently, global challenges such as deforestation, indefensible cultivation practices, loss of flora and fauna, greater risk of climate alteration, as well as the intensification of food shortages, poverty and malnutrition, can be observed and investigated [1]. Proper selection of trees and crop species helps meet wood demand, increase yield, soil fertility, promote sustainability and adequate efficiency of natural resources [2]. The cultivation of cocoa interspersed with tree species and high-density short rotation can be the best option to meet the growing requirements of raw materials for food and industry through the sustainability of natural resources [3].

The agroforestry systems of cacao in Tabasco (Mexico) is an ecological agricultural alternative, economically feasible, sustainable, and conservation of natural resources in the simultaneous preservation of forests and jungles. Given that *Theobroma cacao* is a shade-tolerant plant, this

Figure 1. The cacao agroforestry systems in Tabasco (México).

production system manages genetic diversity of ecosystems in southern Mexico, Tabasco [4]. The objective of this review is to conduct a current investigation of the state of the art of cocoa agrosystems in the state of Tabasco, Mexico.

1.1. *Theobroma cacao* L

The *T. cacao* is a member of the Malvaceae family and is a neotropical species that originated in southern and central Mexico. It is widely grown in more than four countries in the humid tropical regions (Tabasco, Chiapas, Oaxaca, and Veracruz). Cocoa beans are used for chocolate, sweets, and drinks. The cacao pods (fruits) grow mainly on the tree trunk. Photo assimilates are translocated from source organs (leaves) to sink organs (pods) through a long-distance phloem pathway.

Based on the morphological characteristics and geographical origins, three main genetic groups have traditionally been defined within cocoa: Criollo, Forastero, and Trinitario. Trinitario has been recognized as a hybrid "Criollo × Forastero." Recently, molecular markers and chromosome analysis have been used to classify cocoa germplasm into 10 main groups: Amelonado, Contamana, Curaray, Guayana, Iquitos, Marañón, Nanay, Purús, Criollo, and Nacional [5]. The genomes of Criollo and Amelonado have been sequenced and cover 76 and 92% of the estimated genome size, respectively [5, 6].

These genomes are important for research in genetic characterization, phylogenetics, and viability. Tabasco is a state of the Mexican Republic located in the southeastern part of the country (Mexico), with 17 municipalities of which three are the largest producers of cocoa with the agroforestry system: Comalcalco, Cunduacán, and Cárdenas (**Figure 1**). It has a tropical humidity climate, with temperatures ranging from a minimum of 17°C to a maximum of 42°C, with a rainfall of 2000 mm per year and a relative humidity of 90%. The topography is generally flat and low and is covered to a large extent with lakes, lagoons, and wetlands [4]. These biotic and abiotic conditions have favored the agroforestry systems in the production of cocoa of the three groups Criollo, Forastero, and Trinitario, at the same time, an increase in the fertility of the soil, micro- and macroorganisms, conservation and together with the sustainability.

2. The cacao agrosystems

Diversification, efficiency in land use, climate change, sustainability, and permanent vegetation cover are important elements of agroforestry systems, which play an important role in adapting to climate change [7–10]. Agrosystem has been recognized as a process in agriculture as an innovative and promising way to reduce the concentration of atmospheric CO_2 by replacing fossil fuels in the Kyoto Protocol [11].

Tabasco is the first producer of cocoa under agroforestry systems, Mexico. It is mentioned as a green and economically viable alternative with the concurrent conservation of forest remains and economic services (**Table 1**). In this state, two typical systems of cocoa

State	Surfaces (Ha^{-1})		Production (T^{-1})	Performance (T/H^{-1})
	Seeded	Harvested		
Tabasco	40,887	40,668	6995	0.172
Chiapas	16,782	15,280	3947	0.259
Guerrero	250	79	24	0.307

Source: https://www.gob.mx/siap [12].

Table 1. The producing states of *Theobroma cacao* in Mexico under agroforestry system [12].

production are managed: (1) traditional farming system, where cocoa plantations are implanted under forests or natural forests, and herbaceous, shrubby, and upper canopy individuals are eliminated to provide greater light input [13–15]; and (2) cocoa plantations are established in areas where the entire forest or native has been removed and replaced by a single forest species and the cocoa plants are shaded with banana and other crops such as cassava and corn as a provisional shade until forest species provide sufficient shade [13, 16, 17].

Environmental performance in cocoa production as monoculture and agroforestry through life cycle assessment based on ISO 14040 and 14044, with adaptation for local impact indicators. The analysis considered cocoa production at the farm level, from field establishment to the limit of the cocoa crop. The results showed that agroforestry has the smallest contribution to the categories of global impact of global warming, acidification, and eutrophication, which represent carbon dioxide (3.67E+01 kg CO_2^-), sulfur dioxide (4.31E−02 kg SO_2^-), and phosphate (2.25E−05 kg PO_4^-), respectively [18].

Agroforestry with cacao and monoculture of cocoa also had the highest amount of organic carbon and organic matter in the soil, conditions that favor the growth and activity of beneficial microorganisms for the soil rhizosphere (*Pseudomonas* sp. and *Trichoderma* sp.) [19]. In addition, this cocoa agroforestry system had the highest value at 1.36 tons per hectare, as compared to the monoculture of cocoa that was below 1 ton per hectare. Therefore, this agroforestry system is a wise option to promote the environmental sustainability of cocoa crops associated with plant diversity [19].

Agroforestry is increasingly considered as an important adaptation and mitigation strategy against climate change. In particular, the use of fruit trees, trees, and shrubs has been promoted as a practice that contributes to improving soil fertility through nitrogen fixation and the use of macro- and beneficial microorganisms, by increasing the supply of nutrients for the production of crops. While much of the evidence on the impact of tree diversity is based on ranch experiments and correlation analysis, there is a paucity of rigorous evidence under the real conditions of small farmers. Research on the impacts of the adoption and the change of arboreal vegetation such as *Gliricidia sepium* and *Faidherbia albida* [19].

The producers in Malawi adopted the agroforestry system with cocoa and the changes of trees or shrubs of the legume type showed the increase in value to the different crops by 35%. It also had an impact on the disaggregation through stratification by land ownership revealing that farmers with smaller farms of up to 2 acres obtain the greatest benefits. In addition, the use of legume-type shrubs along with corn also significantly increased the value of food crops. This study offers preliminary ideas that contribute to an emerging field of research on the quantitative assessment of agricultural interventions, such as agroforestry practices using new analytical approaches. We provide some policy ideas and recommend the need to design future research on development initiatives that consider small-scale variation in the social, economic, and ecological context of farmers to improve their absorption and adaptation in order to take advantage of the full potential of the agroforestry to improve soil fertility and family food security [20].

2.1. The cocoa beans

Chocolate has been attributed to attainment of optimal human health and development due to its high content of flavonoids that are crucial in reducing the risk or delaying the development of cardiovascular disease, cancer, and other age-related diseases [21].

The production of chocolate had a great impact with the appearance of the diseases of *"Phytophthora," "Ceratocystis fimbriata," "Crinipellis perniciosa,"* and *"Moniliophthora roreri"* in the cocoa plants (**Figure 2**). To recover the production of cocoa, many hybrid plants resistant to diseases have been developed (**Figure 3**). However, some different cocoa hybrids produce

Figure 2. The healthy and diseased *"Moniliophthora roreri"* cocoa pod present in agroforestry systems.

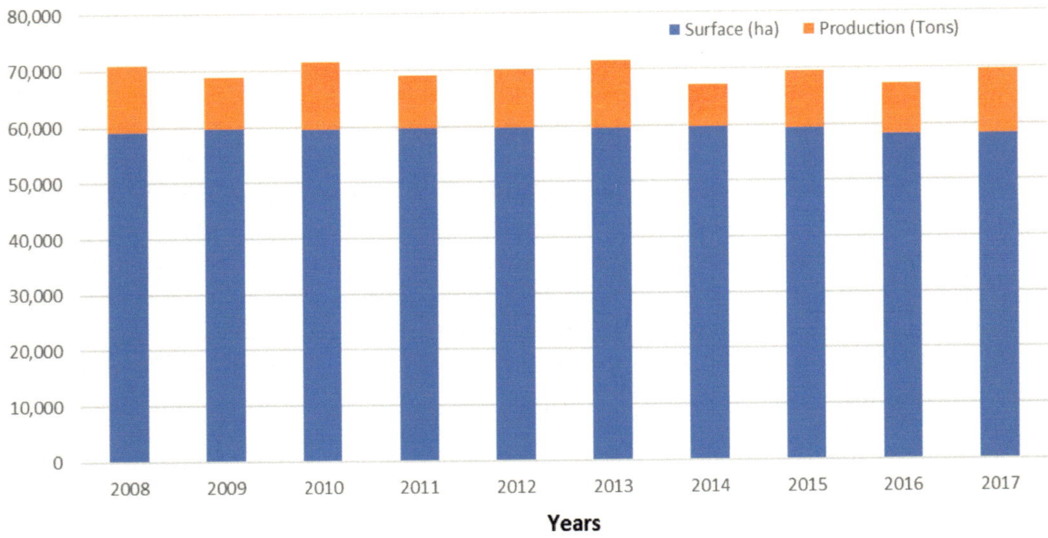

Figure 3. The surface area and production of cocoa in the state of Tabasco [21].

cocoa beans that generate chocolate with variable quality. The fermentation of cocoa beans is a microbiological process that can be applied to the production of chocolate flavor precursors, which leads to overcome the problem of the variable quality of chocolate [22]. The beneficial microorganisms in the use of cocoa production have influence on the microbial communities present in the fermentation process of the compounds involved during the fermentation and the sensory characterization of chocolate. In the bitter or semibitter taste, the use of microorganisms such as: *Saccharomyces cerevisiae*, *Lactobacillus plantarum*, and *Acetobacter pasteurianus* is proposed as starter cultures for the fermentation of cocoa.

Climate change can create serious problems for farmers by increasing the variability of rainfall, droughts, and floods. Understanding how to build the resilience of livelihoods for these purposes is a hallucinatory necessity. Agroforestry can be considered as a potential solution, although many people intuitively link agrosystems with livelihoods against floods and drought, there is little comprehensive empirical evidence. Agrosystems for small farmers can develop livelihood capacities for floods and drought: it is possible to adapt various strategies to climate change, but is crucial for small farmers, and agroecosystem can be a promising option to balance the main production of cocoa, which is mainly chocolate grains [23].

3. Soil chemical properties and nutrient

The soil of the agroforestry systems is enriched with the addition of litter in large quantities by the different plant species, which ultimately improves fertility in terms of organic carbon available from the soil to the plant and the participation of macro- and microorganisms in the soil. The processes of crushing, degrading, and transforming the available nutrients (N, P, and K) provide alternative sources of income and employment to the population of the rural areas that help in the economic systems [13, 24].

The fall of litter is an important input for the replacement of microbial substances in the soil and is one of the most essential ways to maintain soil fertility and quality of vegetable production with free of agrochemicals. The agroforestry system is a proven system of land use to vertically improve the health, quality, and microbial stability against unsuitable climatic conditions. This system contributes to increase the nutritional balance in the soil compared to the open system, that is, pH (7.9), EC (0.43 dSm^{-1}), available nitrogen (253.48 kg/ha), potassium (219.63 kg/ha), organic carbon (1.07%), and available soil phosphorus (22.72 kg/ha) [13, 24].

The spatial integration of land use for the management of soil fertility, soil health, economic services, crop quality, and the key factors that influence the integration of land conservation based on multipurpose agrosystems such as outputs and inputs in the process of agroforestry-cocoa integration and the use of agroforestry space lands have been discussed [13, 25–28].

Author details

Hortensia Brito-Vega*, Edmundo Gómez-Méndez and José Manuel Salaya-Domínguez

*Address all correspondence to: hortensia.brito@ujat.mx

División Académica de Ciencias Agropecuarias, Universidad Juárez Autónoma de Tabasco, Teapa, Tabasco, Mexico

References

[1] Garrity DP. Agroforestry and the achievement of the millennium development goals. Agroforestry Systems. 2004;**61**:5-17

[2] Sharma P, Rai SC, Sharma R, Sharma E. Effects of land use change on soil microbial C, N and P in a Himalayan watershed. Pedobiologia. 2004;**48**:83-92

[3] Sarvade S, Mishra HS, Kaushal R, Chaturvedi S, Tewari S. Wheat (*Triticum aestivum* L.) yield and soil properties as influenced by different Agri-silviculture systems of Terai region, northern India. International Journal of Bio-resource and Stress Management. 2014;**5**(3):350-355

[4] INEGI. Instituto Nacional De Estadistica Y Geografia. 2018. Available from: http://www.inegi.org.mx/geo/contenidos/imgpercepcion/imgsatelite/estacion_recepcion_eris.aspx

[5] Li F, Wu B, Qin X, Yan L, Lai J. Molecular cloning and expression analysis of the sucrose transporter gene family from *Theobroma cacao* L. Gene. 2014;**546**(2):336-341. DOI: 10.1016/j.gene.2014.05.056

[6] Rani S, Chauhan SK, Kumar R, Dhatt KK. Bioeconomic appraisal of flowering annuals for seed production under poplar (*Populus deltoides*) based agroforestry system. Tropical Agricultural Research. 2011;**22**(2):125-133

[7] Chauhan SK, Brar MS, Sharma R. Performance of poplar (*Populus deltoides* Bartr. Ex Marsh) and its effect on wheat yield under agroforestry system in irrigated agro-ecosystem, India. Caspian Journal of Environmental Sciences. 2012;**10**(1):53-60

[8] Chauhan SK, Sharma R, Singh B, Sharma SC. Biomass production, carbon sequestration and economics of on-farm poplar plantations in Punjab, India. Journal of Applied and Natural Science. 2015;**7**(1):452-458

[9] Singh B, Sharma KN. Tree growth and nutrient status of soil in a poplar (*Populus deltoides* Bartr.) based agroforestry system in Punjab, India. Agroforestry Systems. 2007;**70**:125-134

[10] Singh RB. Environmental consequences of agricultural development: A case study from the green revolution state of Haryana, India. Agriculture, Ecosystems and Environment. 2000;**82**:97-103

[11] Ajit HAK, Dhyani SK, Bhat GM, Malik AR, Dutt V, Jain A. Quantification of carbon stocks and sequestration potential through existing agroforestry systems in the hilly Kupwara district of Kashmir valley in India. Current Science. 2017;**113**(4):782-785. DOI: 10.18520/cs/v113/i04/782-785

[12] SIAP. Agrifood and Fisheries Information Service. CACAO. 2018. Available from: http://infosiap.siap.gob.mx:8080/agricola_siap_gobmx/AvanceNacionalCultivo.do

[13] Brito-Vega H, Salaya-Dominguez JM, Gomez-Mendez E, Gomez-Vazquez A, Antele-Gomez JB. Physico-chemical properties of soil and pods (*Theobroma cacao* L.) in cocoa agroforestry systems. Journal of Agronomy. 2018;**17**(1):48-55

[14] Fontes AG, Gama-Rodrigues AC, Gama-Rodrigues EF, Sales MVS, Costa MG, Machado RCR. Nutrient stocks in litterfall and litter in cocoa agroforests in Brazil. Plant and Soil. 2014;**383**:313-335

[15] Monroe PHM, Gama-Rodrigues EF, Gama-Rodrigues AC, Marques JRB. Soil carbon stocks and origin under different cacao agroforestry systems in Southern Bahia, Brazil. Agriculture, Ecosystems and Environment. 2016;**221**:99-108. DOI: 10.1016/j.agee.2016.01.022

[16] Bargali SS. Weight loss and nitrogen release in decomposing wood litter in an age series of eucalypt plantation. Soil Biology and Biochemistry. 1996;**28**:699-702

[17] Bargali SS, Singh SP, Singh RP. Pattern of weight loss and nutrient release in decomposing leaf litter in an age series of eucalypt plantations. Soil Biology and Biochemistry. 1993;**25**:1731-1738

[18] Oliveira PHG, Gama-Rodrigues AC, Gama-Rodrigues EF, Sales MVS. Litter and soil-related variation in functional group abundances in cacao agroforests using structural equation modeling. Ecological Indicators. 2018;**84**:254-262. DOI: 10.1016/j.ecolind.2017.08.030

[19] López-Ferrer ÚC, Brito-Vega H, López-Morales D, Salaya-Domínguez JM, Gómez-Méndez E. *Trichoderma* role in agroforestry-cacaotal systems as an antagonal agent.

Tropical and Subtropical Agroecosystems. 2017;**20**(1):91-100 Available from: http://www.revista.ccba.uady.mx/ojs/index.php/TSA/article/view/2278/1054

[20] Coulibaly JY, Chiputwa B, Nakelse T, Kundhlande G. Adoption of agroforestry and the impact on household food security among farmers in Malawi. Agricultural Systems. 2017:15552-15569. DOI: 10.1016/j.agsy.2017.03.017

[21] Arévalo-Gardini E, Arévalo-Hernández CO, Baligar VC, He ZL. Heavy metal accumulation in leaves and beans of cacao (*Theobroma cacao* L.) in major cacao growing regions in Peru. Science of the Total Environment. 2017;**605-606**:792-800. DOI: 10.1016/j.scitotenv.2017.06.122

[22] da Veiga Moreira IM, de Figueiredo Vilela L, da Cruz Pedroso Miguel MG, Santos C, Lima N, Freitas SR. Impact of a microbial cocktail used as a starter culture on cocoa fermentation and chocolate flavor. Molecules. 2017;**22**(5):1-15. DOI: 10.3390/molecules22050766

[23] Quandt A, Neufeldt H, McCabe JT. The role of agroforestry in building livelihood resilience to floods and drought in semiarid Kenya. Ecology and Society. 2017;**22**(3):1-12. DOI: 10.5751/ES-09461-220310

[24] Sirohi C, Bangarwa KS. Effect of different spacings of poplar-based agroforestry system on soil chemical properties and nutrient status in Haryana, India. Current Science (00113891). 2017;**113**(7):1403-1407. DOI: 10.18520/cs/v113/i07/1403-1407

[25] Guteta D, Abegaz A. Factors influencing scaling up of agroforestry-based spatial land-use integration for soil fertility management in Arsamma watershed, Southwestern Ethiopian Highlands. Journal of Environmental Planning and Management. 2016;**59**(10):1795-1812. DOI: 10.1080/09640568.2015.1090960

[26] Pandey DN. Multifunctional agroforestry systems in India. Current Science. 2007;**92**:455-463

[27] Bisht N, Sah VK, Satyawali K, Tiwari S. Comparison of wheat yield and soil properties under open and poplar based agroforestry system. Journal of Applied and Natural Science. 2017;**9**(3):1540-1543

[28] Puri S, Nair PKR. Agroforestry research for development in India: 25 years of experiences of a national program. Agroforestry Systems. 2004;**61**:437-452